Called Up, Sent Down

THE BEVIN BOYS' WAR

TOM HICKMAN

First published in 2008
This edition published in 2010

The History Press
The Mill, Brimscombe Port
Stroud, Gloucestershire, GL5 2QG
www.thehistorypress.co.uk

British Library Cataloguing in Publication Data.
A catalogue record for this book is available from the British Library.

ISBN 978 0 7524 5749 9

Typesetting and origination by The History Press
Printed in Great Britain

Contents

One

Unlucky Dip

For tens of thousands of boys in Britain the Second World War was an adventure. There was confusion and anxiety as families broke up and fathers and older brothers went into the forces; there was the upheaval of evacuation; and danger when the bombing began. But across the country boys too young to join up watched excitedly as RAF fighters attacked the German bombers overhead. They saw the bombs drop and at night the fires they caused, listened to the pounding of the guns and picked pieces of shrapnel from the rubble of damaged buildings as souvenirs.

As Boy Scouts they packed gasmasks into cardboard boxes, collected waste paper for the war effort, and delivered sandbags to homes. They volunteered for fire-watching, were taught how to tackle incendiaries, acted as messengers for the Civil Defence. They joined one or other of the cadet corps, drilled, went on manoeuvres, fired rifles or even artillery pieces on the shooting range, learnt Morse code and other skills including aircraft recognition. Some joined the Royal Observer Corps, which plotted incoming and outgoing aircraft, others the Home Guard, taking turns of duty on week nights and weekends at munitions plants, installations and aerodromes. Many of those whose education ended at fourteen (even thirteen where schools were destroyed) went to work in the war factories. And almost without exception they entertained the same hope: that the war would last until they were old enough to get into the fight.

It was possible to volunteer for the forces at seventeen and a half and some did. Those who waited for call-up went for medical and interview just before their eighteenth birthday. In December 1943 thousands were called to register, as millions in the four years of conflict had been before them. Those passed fit were seen by a recruiting officer, marked down for army, navy or air force, and went home to await their instructions to report. Days before Christmas the expected OHMS buff envelopes arrived.

For hundreds, they contained news that stunned them. In the next seventeen months over 20,000 other conscripts like them would be similarly stunned. Instead of a uniform they were to get a miner's lamp. Ernest Bevin, the Minister of Labour and National Service, was sending them into the coal pits, where many would remain for up to four years.

*   *   *

Why Bevin felt it necessary to send 20,000 eighteen-year-olds into the country's hardest and most dangerous industry is a tangled tale with its roots in the years between the end of the First World War and the beginning of the Second. These were the years of coalmining's decline, when never less than a quarter of miners were wholly or partly out of work.

Things changed with the declaration of war in 1939. Coal was needed to conduct the war and the miners met the demand for increased productivity. Those who'd served in the forces pre-war and were in the Reserves or the Territorial Army were called up, but their going was balanced from the ranks of the unemployed. Then, eight months later, France fell, Italy entered the conflict on the side of Germany, closing another export market, and things changed again. Nearly a hundred pits shut down virtually overnight and men were again thrown out of work; over 34,000 were idle, with south Wales and the North East most badly affected. Miners' leaders and MPs put strong pressure on the government to drop any restriction on men seeking to work elsewhere and Bevin was persuaded against his better judgement. He raised the age of reservation[1] for military service in the industry from eighteen to thirty, which allowed thousands to leave for the mushrooming munitions factories and construction.

In the early months of 1941 it became apparent that this was a serious mistake.

In 1939 there had been 773,000 miners; now that number had dipped under 700,000 – and the pits were losing as many as 28,000 men a year through retirement, death, long-term illness and disablement. Moreover, the recruitment of fourteen-year-olds to the mining industry was at such a low ebb it wasn't keeping pace with natural wastage. In May, with the shipyards, the steel mills, the power stations and everyone else crying out for ever more coal, an incredulous House of Commons heard for the first time that Britain, a country whose reputation was built on coal, a country used to an abundance of coal, was facing a coal shortage. Total production in the previous year had fallen by 7 million tons from 1939's figure of 231 million. And was still falling.

From the time Churchill came to office, an Essential Work Order applied across industry to prevent workers quitting their jobs and employers from sacking them. Mining was the exception. The strong lobby of mine owners resisted it and there was uncertainty about how the miners, still bitter after two decades of hardship, might react. Though in a sense he was shutting the stable door, Bevin now imposed the Order. And not without misgivings, because it bound the miners to the owners – and the feud of the inter-war years was not forgotten. The miners accepted the control because Bevin had a carrot as well as a stick: the Order gave them, as it did all workers to whom it applied, a guaranteed week – a right they'd never been able to wring from their masters. He also threw in an extra shilling a shift.

The following month Bevin broadcast an appeal for men to return to the pits of their own accord. As those in the munitions factories were now earning more than the mining average of £4 (plus 2s 9d worth of domestic coal) and more again in the shipyards (£5 16s 7d) or in aircraft construction (£6 7s 5d), and in infinitely better conditions, it was hardly surprising that fewer than 500 answered the call. Bevin was obliged to tighten the screw. In July he ordered all ex-miners who'd worked in the industry since 1935 to register; gradually, reluctantly, men were returned to the pits, including from the forces. 'A good many have been brought back, but a good many others don't want to return – they prefer the Army life,' wrote the journalist and broadcaster James Lansdale Hodson. 'N.C.O.'s cannot be forced back, and one hears of the creation for protective purposes of additional "lance-corporals unpaid"'.[2]

Again and again Bevin pleaded that a workforce of 720,000 was critical, but still he couldn't get it. Numbers were reasonably stable at something over 700,000, but production continued to slide: the 224 million tons of 1940 became 206 million in 1941. There were many reasons. Depleted of younger men, especially at the coalface, the miners were beginning to suffer fatigue. They were angered by amendments to the Essential Work Order that allowed them to be prosecuted for absenteeism – which had become persistent among a minority. Above all, they were aggrieved that, even with the increase of 1941, their pay packets, in the most essential of all industries, were still below those in the munitions factories.

Levels of absenteeism increased. In January 1942 men at Betteshanger colliery in Kent struck over the level of allowances for working difficult seams. Under wartime legislation all strikes were illegal and the Ministry of Labour decided to prosecute; over a thousand miners were fined £1, the fifty working the difficult seams £3. Three local union officers were imprisoned. Betteshanger stayed out and other pits came out in sympathy. The Home Secretary released the officials. By May only nine miners had paid their fines; the majority never paid.

In May and June more widespread strikes – 160,000 men out. Worried, the War Cabinet swiftly did three things. It set up a Ministry of Fuel and Power to coordinate coal output and try to increase productivity, improve relations between owners and miners, and persuade the population as a whole to be fuel conscious. With a nod towards nationalisation the pits were put under dual control: government regional controllers to direct operations, but the business of drawing coal left to the owners. A pay settlement at least tempered the miners' dissatisfaction and halted the unrest. The Mineworkers' Federation had demanded an increase of four shillings a shift; they got two and six. Perhaps more importantly, the miners were given something else that the owners had always refused: a minimum national wage.[3]

During the last six months of 1942 output rose; then, at Christmas, fell. The yearly total came in at 203 million tons. There was a continuous manpower problem. Bevin sought non-miner servicemen stationed on home soil willing to transfer into the pits; in September he made mining an option for those registering for military service, but by the end of the year no more than 1,100 elected to take it. Early in 1943 some collieries in Lancashire and Cheshire closed because of a scarcity of able-bodied men.

The old anxieties escalated in the first six months of the year. More absenteeism. More strikes. Between 2 July and 13 November there were 421 stoppages due to disputes. Miners at Valleyfield colliery in Fife, and Cortonwood and Hatfield Main near Doncaster were fined; in Nottinghamshire the imprisonment of an eighteen-year-old surface worker for refusing to work underground brought out 15,000, resulting in the loss of 50,000 tons of coal.

By July the Ministry of Labour had carried out a different calculation of the workforce and it made grim reading. When legitimate absenteeism through sickness and injury occurring for the whole of any week was added to wilful absenteeism and the total deducted from the numbers on the books, the *effective* labour strength was 646,600. Bevin told the miners' conference in Blackpool:

At the end of this coal year there won't be enough men or boys in the industry to carry it on . . . . Every bit of territory we take from the enemy we have to find coal for . . . I will have to resort to some desperate remedies during the coming year . . . I shall have to direct young men to you.

His disclosure created furore. There was already talk of new conscripts from coalmining areas having pressure put on them to opt for the pits; now the miners' Federation asked did the Minister mean to direct only young men from mining areas? And was this another device to evade the issue of nationalising the pits, which was the only way to put right coalmining's ills? This wasn't the first time in the war that the age-old question of the State taking over the collieries had raised its head (at its Ayrshire conference in July 1941 the Federation had reiterated its demand for State control). Here, however, it was tangled up with a different emotive issue. The *Sunday Pictorial* wrote an open letter to Bevin demanding 'no conscription without nationalisation'. Bevin answered the Federation's first point robustly, saying that if he was driven to using compulsory powers he would apply them equally to all classes and areas. On the subject of nationalisation he stayed silent.

All over the country miners were being fined for taking part in illegal strikes or for absenteeism, and in September, sixteen in Lanarkshire briefly went to prison for refusing to pay fines. 'The mining industry is developing into a raging maelstrom of discontent,' argued Seymour Cocks, the Labour MP for Broxtowe in Nottinghamshire, supporting the call for nationalisation during a three-day parliamentary debate on coal.

Churchill dismissed even consideration of the issue until the war was won:

I hold the opinion that there is nothing in the present coal situation which would justify the violent overturn of our present system. Even if the overturn were well conceived, which is improbable having regard to the hurried conditions in which it would be born, it would cause more trouble than it was worth and the reaction engendered might be deeply harmful to our war effort, and might well prolong the war.

As far as the miners were concerned, he chose to be expedient:

We are told of the great unrest in the mining industry. I think that it is a little unjust to the miners . . . The loss by stoppages compares very favourably with the last war . . . It must be remembered that we are in our fifth year of war. There is a fifth war year mentality . . . I cannot see anything in the mining situation which makes me apprehend that this will be found to be the one gloomy failure in our national struggle . . .

I can well realise that anxiety exists among the miners about what is to happen to them and their industry after the war. They had a very grim experience after the last war which went on biting at them . . . We can all lie awake thinking of the nightmares that we are going to suffer after the war is over and everyone has perplexities and anxieties. But I, being an optimist, do not think the peace is going to be so bad as the war.

That put the lid on nationalisation for the time being, but it was an issue that never went away.

In the same debate Gwilym Lloyd George, the Minister of Fuel and Power, laid down another marker for call-up to the mines, but for weeks Bevin held off, hoping if not expecting to find more men from somewhere. In the previous two years over 60,000 miners had been returned to the pits – 48,890 from industry, 9,600 from the army, 1,800 from the RAF. A final comb-out among older servicemen stationed in Britain in units not earmarked for forthcoming operations winkled out another 7,000. At this juncture only 3,366 'optants' – conscripts opting on registration – had signed up, as had only 3,500 from the forces, still leaving Bevin short of his 720,000 objective (an objective, in fact, that would never be reached). The War Cabinet considered, and rejected, bringing back miners from units overseas and dismissed as unworkable the possibility of using the hundreds of miners among German and Italian prisoners in the country. That idea had occurred to others including Daniel Davies from Aberdare in Glamorgan, who in September had written to *The Times*: 'North Italian miners are known to be first class, and the Germans are used to difficult and dangerous seams. These men, who must be fed and supervised, could produce coal and so hasten the day of their release.' One Tory MP suggested convicted felons be offered prison or the pits, which did not go down well in the mining communities.

As part of an ongoing campaign, the Ministry of Labour was sending out letters to those reaching call-up age, seeking anyone prepared to put himself forward. The letter might have urged recipients to an act or patriotism, but it read like a job application form: 'If you desire to be considered for underground mining work, you should complete the space below and return this form without delay.' Few, if any, desired. In early winter Bevin made a radio broadcast to sixth-formers in grammar and public schools that did appeal to their patriotism; and he promised that any who made the choice would not be held in mining after the war, adding: '[But] some might like to stay; the industry is undergoing a revolution with the development of mechanisation – there will be a need for more technicians.' He was as unsuccessful as the Ministry's letter.

Unrest continued to erupt in the coalfields, with considerable loss of output. As plans advanced for the Second Front, with prodigious implications for the mining industry, and domestic use of fuel surged at the beginning of the second consecutive bad winter of the war (and on *that* front there were warning ripples about fuel stocks, with a hundred instances of firms stopping production for short periods because they had run out of coal), Bevin bowed to the inevitable: compulsory conscription to the pits it was going to be. On 2 December in a statement to the House he explained how he planned to conduct it. The selection, he said, would be made from men born on or after 1 January 1918 (that is, up to twenty-five years of age) who were medically grade I (or II 'if their disability is foot defects only'). The only exemptions from it were men accepted for flying duties in the RAF or Fleet Air Arm, for service in submarines, and for bomb disposal; and men in certain highly skilled trades of great value to the armed services – such men already barred from being optants. For selection to be fair and seen to be fair, he proposed:

> to resort to the most impartial method of all, that of the ballot. A draw will be made from time to time of one or more of the figures from 0 to 9 and those men whose National Service Registration Certificate numbers happen to end with the figure or figures thus drawn will be transferred to coalmining.

The first draw took place on 14 December – the ballot resorted to for the first time since the second half of the eighteenth century when the militia was raised from parish lists. Until the end of the war in Europe the scheme would take one conscript in ten, but so badly were men needed in the pits that two draws

were made that day – and were on six of the later thirty-two dates that followed
– thereby claiming one in five.

The newspapers reported that the draw took place in the presence of Bevin,
Lloyd George and Rab Butler, President of the Board of Trade, and was made
by a junior member of staff. More colourfully, the story subsequently changed:
the numbers were plucked from Bevin's homburg by his secretary whose name,
a Ministry spokesman was quoted as saying, 'is concealed lest she should be
molested by mothers of boys who were sent to the coalmines.' Perhaps Bevin
did resort to this little ritual, though he never said so and perhaps it owes more
to imagination than fact. On occasion Bevin was as likely to have asked another
member of staff, or dipped a hand into his own hat. Or as any arbitrary method
was as fair, or unfair, as another, he might simply have thought of a number off
the top of his head.

At the end of the 1943 coal year, output was still plummeting: 194
million tons.

*   *   *

The possibility that some conscripts were going to be directed down the pits was
on the cards from summer 1942, when Bevin addressed the miners in Blackpool,
and became a certainty after the October coal debate. Despite the considerable
publicity some soon-to-be conscripts remained unaware; few of those who were
aware thought it could happen to them.

Numbers drawn in the ballot were never revealed during the war.[4] The first,
therefore, that men knew that the finger of fate was pointing at them was when
their call-up papers dropped on the mat. The first batch arrived just before
Christmas 1943 – an unwanted Christmas present. Geoff Baker, son of a London
Docklands parson who'd just left boarding school and was waiting to go into the
Royal Marines, reacted with disbelief:

> This is a load of rubbish, I thought. It used the word 'selected', as if I was
> privileged – I'll never forget that word. It was almost more than I could bear.
> I put the thing in a drawer, wildly thinking that with a bit of luck anything
> relating to me would be lost in the Christmas rush. Alas. For a lad whose
> school motto was Quae sursum sunt quaerite [Seek those things which are
> above] I seemed to be going in the wrong direction.

In some cases the dashing of hopes was even more painful. Their papers ordered
some men to report to the armed service they'd expected – only for them to be
countermanded immediately by others. One day in June 1944 Roy Doorbar, a
butcher from the village of Smallthorne near Stoke-on-Trent, was to be on his way to
Litchfield barracks and the North Staffs Regiment; the very next he was a coalminer:

I was the only one left at home out of four brothers, two already in the forces, and I wanted to kill Germans, didn't I? My father, who'd died earlier in the war, was a miner in the twenties and been badly hurt. He'd become an insurance agent and swore no son of his would ever go down the pit. A good job I've a sense of humour.

Two or three months into the scheme, intakes of conscripts were much more likely to know all about 'Bevin's tombola' and the anxious wait between registration and call-up. Knowing that they might be unlucky didn't make it any easier to take for those who were. David Reekie from Deptford, an accounts clerk in a comic and magazine publisher's in central London, chosen to be a navy wireless telegraphist, 'was in a state of catatonic shock for days'. 'It was a bigger bombshell than anything the Germans managed,' for Ian McInnes, and he'd experienced bombing in London while taking an engineering crash course in anticipation of a commission in REME, and the arrival of the VI rockets both there and his home town of Dover, where he'd also witnessed the cross-Channel shelling. Warwick Taylor, from Harrow, a junior clerk in a City importers/exporters faced no anxious wait to discover what would happen to him – he found out when he registered at Ruislip. Registration boards generally didn't know whether men they saw had or would draw the short straw (the delay between the ballot and the board, which didn't always come in that order, was usually brief); and if they did know they kept quiet about it.

But there was this chap there from the Ministry of Labour who told me: 'Hard luck, chum, you've been nabbed for the coalmines.' I said, 'Ridiculous, I'm going in the RAF.' 'Oh, no you're not,' he said. It was like a slap in the face – three-and-a-half years in the ATC counted for nothing.

Morry Pearce, who worked in his father's bakery and grocery store in Monk Sherbourne, near Basingstoke, perhaps had even more reason for feeling aggrieved. A little after reaching seventeen and a half he'd volunteered for the Marines:

There were a dozen of us at the temporary recruiting office in Aldershot. We had a medical, did a written exam, some mental arithmetic and answered some questions, to see if you had anything up here. Four of us were kept back and the others let go. 'You are now Royal Marines,' the recruiting sergeant said. 'Go home and wait for further orders.' I expected something in a fortnight. After a month or more when nothing came I went on my motorbike to the permanent Marine office in Reading. The officer took me through to the back and told me, 'We can't take you, son.' I think he must have had the result of the ballot but he didn't let on. I was devastated. From when I was small I admired the Marines. If I was going to put my life on the line in action I wanted the best fighting men in the world around me to get me through. But I'd been overtaken by my registration group getting called.

The shock their papers gave them turned to anger for many. Typical was Dave Moody, who worked on his father's smallholding in the Hampshire village of Lockerley, a Home Guard from fourteen and 'proud to be parading with my rifle, ten rounds of ammunition and a bayonet', due to join the county regiment. 'I had two cousins in the army, two in the air force – and I was to be left out, sort of thing. I had a few things to say at the labour exchange in Romsey. They practically kicked me out.' They practically kicked out Les Thomas at his labour exchange too. At fourteen he had become a telegram boy 'running round London in the air raids'. Now he was the company cashier of a fountain pen company near his home in Hackney, and he

tore down and threw my papers over the counter. At work next day a telegram arrived telling me to report back to the Exchange or I'd be subject to arrest. The manager talked me round, he said I'd upset my mother.

But I was bloody upset. I'd taken a test for the navy, rigorous, you had to get 55 of 60, otherwise it was the army. I was offered training as a naval officer in decoding. And my mother would have got an allowance, being a widow. My brother was four years older and in a reserved occupation as an engineer and he was living at home, but my 15 shillings wages would be missed. The allowance would have made up for that. So I was disappointed for my mother and more disappointed for myself. I'd lost an opportunity that could have changed my whole life.

The call-up notification seemed to hold out some hope of escaping its decision: 'You may appeal against this notification if you consider that there any special circumstances connected with coalmining which would make it an exceptional hardship for you to be employed in this work.' Four in ten men did appeal, the majority of those who were cadets in the genuine belief that their service was being overlooked. They armed themselves with what written evidence they could to appear at their tribunal. Harold Gibson, a junior in the Blackburn tax office, brought a letter from the senior physics master at Queen Elizabeth's grammar who was also the officer in charge of the school ATC squadron:

Corporal Gibson has thrown himself wholeheartedly and diligently into the specialised technical training to fit himself for service in one of the branches of H.M.Forces. He has passed the 1st Class Star test, the proficiency grade and the advanced training examination. This latter is of a very high standard and held by very few cadets in the entire ATC.

His record at school is equally distinguished.

I feel that he would be of greater value to the country in some form of technical service in one of the Armed Forces.

Reasoning that a figure of authority might sway a tribunal, some brought fathers or brothers already in uniform, often with a guarantee from their commanding officer that a place in his unit awaited the appellant. David Roland, a trainee cutter at a children's clothing factory in Hackney, who 'had no objection to being called up but a very strong objection to being called down', took one of his twin sisters, an officer in the ATS, thinking 'she'll have a lot of pulling power'. While some men argued their own case that they were doing essential war work and should be allowed to continue, a few brought their bosses to argue it for them.

For others, the hope of salvation rested on challenging their medical grading. Arthur Gilbert, from the Staffordshire town of Cheadle and just out of grammar school in Uttoxeter, brought verification from his doctor about his asthma:

> It was intermittent and, okay, it didn't stop me playing football and cricket. But it used to hit me maybe once a month – asthma seemed different in those days, and there were no inhalers. Attacks lasted a day or two. And they were that bad it took me ten minutes to go up the stairs.

Derek Thompson from Salford in Lancashire, a diamond tool apprentice in a company making industrial tubes, was convinced he had more than enough evidence that he was medically unfit:

> At seventeen I'd tried and failed three times to get into the navy. People used to come round the school giving talks on the forces and I tried for the officer 'Y' scheme but found it was closed. Then I tried for the Fleet Air Arm and discovered I was colour blind. Then I tried to become an ordinary seaman and was turned down because of defective hearing, though I wasn't aware there was anything wrong with it – but as a child a group of us playing in a field were using elderberry twigs as spears and I'd got one right in the left ear; the inside was all jumbled up.

Although it smacked of desperation, others tried to persuade their tribunal that something physical about them made them unsuitable for what was in store. Morry Pearce decided that claiming to suffer from acute claustrophobia might get him out of it; others argued that they were too tall for coal seams; that they weren't strong enough for manual labour; that as they wore glasses the dust underground would make them a danger to themselves and others.

Virtually without exception appeals were dismissed out of hand. Those with an apparent medical problem got a stay of execution by being referred to a specialist

– who invariably passed them as fit. Derek Thompson's hearing was pronounced fine, but might not have been if he had lied: 'The cursed little man told me to kneel down behind a chair, whispered, and asked me what he'd said. I told him straight out. It never entered my head to say I couldn't hear – I was brought up to be honest.' The majority of appeals that were allowed were on compassionate grounds, but even these were few; at least one that should have been granted was turned down: Frank Pratt from Twickenham, who worked in a factory turning out small parts for Rolls-Royce engines, was the only carer for his mother, who had multiple sclerosis.

> My father was a regular, a company sergeant-major in the Dorset Regiment. He'd been at Dunkirk and was now stationed at a training depot in Suffolk. There was no way he could get back at the time. And I had to make what arrangements I could for my mother and leave her. It was bloody wrong.

In their heart of hearts it wasn't likely that more than a minority really expected to win their appeal. But the way in which almost all of them lost left a burning sense of resentment. For Geoffrey Mockford from Babraham in rural Cambridgeshire, a cabinetmaker whose workplace had gone over to making instrument cases for field telephones and bombsights, who had been accepted by the Fleet Air Arm and was hoping to be repairing aircraft on aircraft carriers,

> the appeal was the greatest farce I had ever experienced. You were allowed five minutes to get into the appeal room, state your case, hear the judgement, and then get out. I tried to explain as clearly as I could that a) I'd already been accepted into the Fleet Air Arm and b) that I didn't think I had any abilities which would be of use in the mines, so it seemed common sense to allow me to do work for which I had some ability. The chairman cut me short. 'Your appeal is rejected.' He didn't even consult the other two members. I think the instruction was to reject all appeals unless some very obvious blunder had been made. I came out feeling quite sick.

'They weren't interested in anything you had to say,' says Alan Lane, one of those who appealed because he wore glasses, 'which I'd done since I was four or five and I didn't think it would be clever for me to work underground wearing them. One of the panel told me "I have heard that poor eyesight improves underground." I'm not kidding. The man on the end, I swear, was asleep.'

'Appeal court? Kangaroo court,' adds Dan White, who had just completed a four-year apprenticeship as a diesel mechanic in Hastings. 'In, say your piece, blah blah, appeal dismissed – it was as quick as that.' To Roland Garratt from Birmingham, who had spent two years making hydraulic pumps for the undercarriages and wing flaps of warplanes, it was 'like facing a firing squad with the bullets already in the breech'. Turned down by his tribunal in Croydon, Dennis Faulkner, an engineer at the local BBC overseas receiving station, asked why. '"We never uphold an appeal," the chairman told me. My next question: "In which case what's the point of holding a tribunal?" The answer: "We live in a democracy."'

Unusually, George Ralston was given a longer hearing by his panel: but his case was out of the ordinary run.

Grandson and son of Scottish miners, and one of twins, he'd come to Corby in Northamptonshire when his father's pit in Lanarkshire, with many others, closed following France's capitulation and he had come south looking for work. He found it in a steel mill making tubes for PLUTO, the pipeline under the ocean (or rather the Channel), which would provide the Allies with petrol in the 1944 invasion of Europe. At sixteen George Ralston joined him in the mill, and a few weeks before he was due to register went with his twin brother Robert to put himself forward.

The case he put to his tribunal 'was that my twin brother and I had always been together and I would like to join him in the army'. Ralston was asked to leave the room while the tribunal deliberated. He was recalled to hear their suggestion – which was 'to ask my brother to join me in the coalmines. And obviously I couldn't ask him to do that.' In splitting up the Ralstons, the Bevin scheme broke with a convention: the forces always tried to keep twins together, if that was what they wanted (a practice that continued right through the post-war years of national service that ended in the early sixties). The Shaffer twins, Peter and Anthony, both to achieve fame as playwrights, were kept together: both were sent into the pits, though only one of them drew the short straw. The late Anthony Shaffer was disgruntled; he had been anticipating intelligence work. The public schools were regularly approached by the heads of the services including the secret services and Shaffer, a pupil at St Paul's (evacuated from London to Berkhamsted in Hertfordshire), was already, as he wrote in his memoir,[5] imagining himself in a world of 'Trenchcoats, turned-up collars and soft-brimmed hats; safe houses, code books and rice paper messages (to be swallowed)'. He had been sent to see 'a certain Colonel X' in a building near Victoria station and done rather badly translating a French railway bill of lading, but the colonel's parting words, 'I expect we will manage to find something for you', made him confident of a glamorous war. That his war was to be spent down Chislet pit in the Kent coalfield Shaffer attributed not to the ballot but the colonel's sense of humour.

To a lot of disappointed and bitter young men the tribunals were insensitive and uncaring. Not only were judgements abrupt, but witnesses were often refused permission to speak. To be fair to the panels, they were all but overwhelmed by numbers. They were being confronted by the same arguments over and over, many of which were trivial in the broad scheme of things. And it was obvious that a proportion of appellants were trying it on. Reg Fisher, a motor mechanic from Wembley, argued he should be spared because he was working for a commercial lorry company with army and RAF contracts. 'Of course it didn't wash, but it was worth a try,' he says. Comments Morry Pearce: 'When I got to the tribunal I realised I had a problem – there were 24 other claustrophobics.' It was regrettable that the prior service training of so many was going to be wasted but even that was irrelevant. The ballot – and the country's need for men to win coal – overrode all considerations.

Numbers of men didn't wait for a tribunal hearing before seeking a way out. Some who had had their original call-up papers rescinded by the ballot rapidly reported to the destinations detailed in their first instructions, claiming a mistake was to blame for their names being missing from any paperwork. George Poston, an engineering undergraduate, went straight from his home in Leicester to the Royal Engineers' training battalion in Wrotham, Kent, and tried to bluff his way. 'But when it was discovered I'd received a communication about the Bevin Boy scheme, REME sent me packing.' Others tried to enlist. In Ian McInnes' case, 'Arnheim had just taken place. I rushed to the recruiting office. "You must need paratroopers, glider pilots, I'll do anything."' Ken Sadler, a salesman in the Co-operative Wholesale Society warehouse in Newcastle-on-Tyne, tried a variety of less predictable escapes:

> Even in the war ads were going in the press for the Palestine Police, the Hong Kong Police, the South African Rifles. I tried them all. I tried to emigrate to Canada or Australia. As I'd worked on Newcastle quayside – as a messenger for a coal company, ironically – I knew all the merchant navy companies. I tried three or four but no one would touch me once I told them about my call-up papers.

That six men did not appeal for every four that did is no indication that they didn't mind. Hardly a man wasn't shocked by his fate. But many took it on the chin out of a sense of obligation. 'I simply accepted it as something that was chosen for me – my duty if you like,' is how Ron Bown, working in a Wolverhampton factory that made batteries and rectifiers for the navy, saw it. 'In those days most young men did as they were told.' Others accepted for the sake of their mothers who had lost an older son, were terrified of losing another, and pleaded that the pits were a safer place than the battlefield. At registration Cecil Kelly, a gardener in Bolton, even opted for that reason: 'My brother Lewis was in the army and killed at 20 in Italy. My mother was traumatised. I did what I did for her.'

Others didn't appeal because they anticipated that the process was largely pointless, like Brian Evans, a junior reporter in Stourbridge in the Midlands. He felt mocked by the timing of his papers' arrival – they were waiting for him when he returned from two weeks at RAF Hixon in Staffordshire, where, 'proudly wearing the white cap flash of the RAF reserve', he'd been doing circuits in a

Wellington bomber practising evasive action from enemy fighters. His mother urged him to appeal on compassionate grounds, 'although there weren't any – my father was very much alive and living with us, working in the timber control office in Birmingham'. He did, however, try evasive action by attempting to enlist in the Reconnaissance Corps ('I rather fancied a scout car'), but failed like everyone who tried the enlisting route. He was realistic: 'Once you were in the bag you were in the bag. But you can imagine what I felt that the war needed me down there and not in the air.'

*   *   *

Ernest Bevin, the Somerset farm boy and Bristol drayman, was made Minister of Labour and National Service by Winston Churchill as soon as he became Prime Minister in May 1940. Churchill considered Bevin to have 'the temperament of a born fighter' – and, as until now he had been the general secretary of Britain's biggest union, the Transport & General Workers, he could carry the unions with him.

Before Bevin arrived at the Ministry, little had been done to mobilise manpower other than for the forces; despite the critical shortage of armaments and equipment, nearly three-quarters of a million workers were unemployed. All that changed. The Emergency Powers Act, passed in a single day, gave Bevin absolute authority over everyone in the country between fourteen and sixty-four (33 million people), making him the most powerful man in the Cabinet after Churchill, and he used that power to mobilise the entire population, with very little friction. Millions were compulsorily directed into essential employment. The trust the unions generally had in Bevin made them accept the loss of cherished privileges; skilled trades allowed the entry of unskilled labour, and under the 'designated craftsman' scheme, skilled men agreed to work when necessary as labourers. In 1941 Bevin's appeal to women to join the industrial army was met with the same sense of cooperation.[6] By 1943, over 8 million women were in paid work, in munitions, in the Land Army, in hundreds of jobs that were previously male preserves – women became train guards and tram drivers, telephone engineers, welders, plumbers, crane operators. Another million were in voluntary jobs.

Like Lord Woolton, the Minister of Food, people trusted Bevin even more than the unions. In their eyes he made only two really unpopular decisions, both in 1943. One was the 'granny call-up', the registration in July of women aged forty-six to fifty for war work. The other in December was to send eighteen-year-old conscripts into the mines.

The Bevin Boy scheme was unpopular with the RAF, too. This letter passed between the Air Ministry and the Ministry of Labour on 20 October 1943, when the details were being worked out:

My dear de Villiers,

I have your letter of the 13th October, in which you discuss the question of implementing the Government decision to call up men for the coal mines as they are now called up for the Forces.

We are anxious, of course, to assist you as much as possible in your difficult task of creating the procedure of selection, and we certainly welcome your proposal to exclude from the ballot the men who have been accepted for flying service. This is undoubtedly sound, in view of the quality required in men for air crew duties. With regard to the exclusion of the unasterisked tradesmen of certain categories, I assume you refer to the occupations, about 60 in number, which we refer to your Department where men apply for other air crew service . . .

I am afraid we must register a strong protest against members of the Air Training Corps being included, having regard to their training in air subjects and trades. To include them and to post a proportion of them for coal mining would be a complete waste of this specialised training. I am sorry that I must press for men with Air Training Corps service to be excluded on the grounds that they have received this training specifically to prepare them for Royal Air Force and Fleet Air Arm service.

I appreciate the points you made on the steps to be taken to prevent a man evading the ballot by volunteering for the forces after registration and subject to the men with A.T.C. service being excluded, we agree that after registration a man will not be accepted for ground service as a volunteer unless he is clear of the ballot . . .

With regard to the pre-registration volunteer, we would certainly wish to prevent the enlistment of a man who volunteers for the Services in order to evade being included later in the ballot, but equally we would not wish to be precluded from taking as volunteers A.T.C. cadets (having six months' enrolled service) or men, though non A.T.C., required for special ground categories, such as Aircraft Apprentices and recruits for Wireless trades . . .
Yours ever
E.H.Donnell

In the October coal debate in the Commons, Seymour Cocks had been emphatic in saying that if conscripts were to become coalminers, 'Eton and Harrow must send their quota as well as the elementary schools', evidently seeing the issue in class terms as had the Minerworkers' Federation.[7] Unsurprisingly, middle-class parents were better able to give voice to their dismay at what was happening to their sons and to enlist the support of headmasters, serving officers, politicians and clergymen. What went unsaid was that the majority of the middle class felt mining was beneath their sons. As did more of their sons than at the time would have been prepared to admit it. Harold Gibson was of that mind:

I lived in a 1932 semi-detached house, I had two distinctions and six credits in my School Certificate, and I worked in a tax office. I'd seen miners when visiting relations in Burnley, not in Blackburn. I was being snobbish but I was brought up Congregational – nothing to do with working-class people. As far as I was concerned coal came in a sack.

But the Bevin Boy scheme was unpopular among the working class, many being as suspicious as the miners' Federation that they were being singled out. Mining communities were particularly incensed. 'My father was disgusted,' says Arthur Gilbert, whose home town of Cheadle was a mining town.

He wasn't in the pits, he was in the copper works, but an uncle of mine was killed underground when I was about seven. Nearly everyone I knew was either in the copper works or in the pits – the wife's father was a pitman, so was her grandfather.

A lot of men like my father vowed their sons would never follow them into the pits. The years before the war were hard. Families struggled to make ends meet. Hungry children would be fed by neighbours who were little better off themselves. Hand-me-down clothes were part of a kid's upbringing, or not having shoes. I remember children begging for food.

I used to hear the miners' clogs clattering by the house and see the local pit from my bedroom window. I saw the blue scars and the surprising number who had missing fingers. I lived in dread of being a miner.

In the Welsh mining village of Cwmdare near Aberdare, Desmond Edwards' father was another disappointed man.

He wanted my brother and me to escape the inevitable life, and the way to do it was through education – education was the entrance to a different life, he used to say. That's why we won scholarships to the country grammar and I'd done a year at Aber [ystwyth] university. When the ballot directed me to the coalmines, it seemed fate had caught up with me.

Edwards had resigned himself to being forced into a job his father, like other miners, had sworn he would never do, but he still went the short distance to Oakwood training centre near Blackwood in a mood of 'angry frustration'. Those attending university during the war were given a clear year before call-up that could come at any time thereafter.[8] Edwards, reading geography and geology, had had almost a full second year and was within six weeks of his second-year exams. He asked for six weeks' grace. The Ministry of Labour refused, 'rendering at a stroke the whole year's course worthless'.

The row over the rightfulness of turning conscripts into miners rumbled on into the early months of 1944. In February, when the first Bevin Boys were finishing their initial training and were being assigned to their pits, Lord Keyes in the House of Lords moved that men who had trained for the forces should be on Bevin's exempt list. That they were not was an 'outrage', he said, adding: 'The ballot system involves a waste of money by the service departments, a waste of time by the splendid team of voluntary officers and instructors in cadet units, and the complete destruction of any faith boys may have had in the guide they received from youth committees.' He found support from Lord Elton, who spoke of 'an alarming falling off in the intake of young men with the necessary intellectual and technical qualifications for some special branches of the services', and urged 'the importance of not withholding from the services potential officers of technical merit.'

It was left to the Earl of Munster, Under-Secretary at the India Office, to remind his fellow peers that the ballot was intended to be fair and equitable. A quarter of those who registered on 11 December were members of the ATC, the Sea Cadet Corps, the Army Cadet Force, Junior Training Corps or Home Guard. If that quarter had been ineligible for mining, he pointed out reasonably, 'To obtain the required number it follows that one-third more of the remaining registrants would have had to be selected and their chances of being selected would have increased by $33^{1}/_{3}$ per cent.'

Lord Keyes withdrew his motion.

A strike sparked at the end of the following month by an apprentice in South Shields finding himself destined for the pits brought out 5,000 shipyard and engineering apprentices on Tyneside and another 10,000 on the Clyde. A thousand aircraft apprentices in Huddersfield joined in, as did a few others in engineering establishments in Lanarkshire.

Back in December, when Bevin announced the ballot, apprentices on the Tyne had formed an unofficial guild to offer collective resistance to it. In February when one of their members had his number drawn from the Minister's hat, they issued an imprudent statement demanding legislation within three weeks that would give unconditional exemption from the ballot for all engineering apprentices. 'We refuse to carry the burden imposed on the industry by the lust for profit and inefficiency of the coal-owners,' the statement said, adding, confusingly: 'Since they are directly responsible for the coal crisis it is against them that compulsion must be directed.'

The priority of shipbuilding and repair, and heavy engineering also, was so great that apprentices whose employers sought deferment for them were automatically granted exemption from call-up, for the forces or otherwise, until they were out of their apprenticeship or reached the age of twenty. The apprentice at the centre of the dispute had not finished his apprenticeship. But his boss, an electrical sub-contractor, had not sought a deferment; and the job on which he was employed – rewiring a building at the time of call-up – could hardly be classified as high priority.

The TUC and the unions told the strikers to return to work. So did Bevin, who made it clear that thousands of non-deferred apprentices had been called up throughout the war, and how conscripts served their country was a matter for the government. The apprentices refused to listen, their leaders uttering defiance ('Bevin won't climb down so we'll pull him down'), their highly charged rhetoric making the Minister of Labour declare they were being manipulated 'by irresponsible mischief-makers with the purpose of coercing the Government' and he began issuing the strikers with notices to present themselves for an armed services medical. *The Times* quoted the strikers as being 'surprised'. Their surprise was such that within two weeks they had all gone back.[9]

*   *   *

For a ballotee determined to stick to his refusal to become a miner, form E.D. 383A Emergency Powers (Defence) Acts, 1939–1941, Direction issued under Regulation 58A of the Defence (General) Regulations, 1939, which was included in his call-up documents, spelt out the alternative:

> Any person failing to comply with a direction under Regulation 58A . . . is liable on summary conviction to imprisonment for a term not exceeding three months, or to a fine not exceeding £100 or to both such imprisonment and such fine. Any person failing to comply after such a conviction is liable on a further conviction to a fine not exceeding five pounds for every day on which the failure continues.

Bert McBain-Lee, a clerk in an insurance office in Liverpool, was too incensed when he got his papers to notice the threat. He had been at college, waiting to go to university to read electrical engineering, but, 'coming from a King and country family and wanting to do my bit', had left and taken a temporary job while he waited to join the army – preferably the Cameron Highlanders, in which his father was an officer, he had told the recruiting sergeant at registration,

> or failing that any Highland regiment or failing that, the Royal Signals – I had semaphore and Morse at respectable speeds. 'You've got it all thought out, lad, haven't you?' he said, and I had. Except I hadn't anticipated my number being drawn in April 1944. When my mother brought the post to the breakfast table the notification just about knocked me off my chair.

Like so many others he 'kicked up hell' at the labour exchange, before rushing off to see a local JP and then his MP. His MP listened to him and said: '"You know

George Martin?" I told him I did, he was a year ahead of me at school and had gone into the navy. "No, he didn't, he's in jail – three months at hard labour for refusing to go down the pit." That drew me up short.'

At David Reekie's tribunal, where his submission to be allowed to become the telegraphist the navy had chosen him to be was abruptly dismissed, the chairman wagged a finger, telling him to do as he was told 'or you must take the alternative'.

I knew what the alternative was, of course. But I was a mummy's boy tied to her apron strings. I was mollycoddled at home – two sisters and various other ladies who were paying guests in my mother's guesthouse. Some brave souls went to prison, but I wasn't brave enough. I cried into my white flag and surrendered.

Laconically he adds: 'Either way I was going to be sent down.'

In the initial blind rage at the cards dealt by blind fate, quite a few ballotees considered prison. The majority calmed down, were talked out of it or had family reasons for complying with their instructions. One was Dan Duhig from Popular, who had been engaged in making submarine periscopes and chain linkages for invasion barges. He was accompanied to his tribunal by his boss 'because he wanted to hold on to me'; but when the appeal failed and Duhig was determined to go to prison, his boss persuaded him against doing so: 'If I went to prison I'd have a prison record. And after the war with the troops home and jobs hard to come by, I wouldn't get one.' Ernie Jefferies, a spectacle-frame maker from Elephant and Castle, London, was another:

I'd dreamed of being in the desert with Montgomery. I might have done something silly, but my mother had died two weeks earlier and my dad was in a state. 'We don't want trouble by bring the police round here,' he said. Under the circumstances I didn't want to go against him.

But some men flatly refused to reconsider. According to a figure given in the House by the Home Secretary, Herbert Morrison, in the first year of the implementation of the ballot – 1944 – 500 of the 15,362 ballotees were prosecuted for disobeying their direction order and 147 of them went to jail then or later after they absconded. There seems to be no existing figure for how many of 1945's 5,534 ballotees were prosecuted or imprisoned.

More might have been locked up if they hadn't been prevented by the police, trying to act in what they thought was their best interests. Syd Walker, until call-up employed in a Birmingham factory that heat-treated armoured plating for tanks, worked down the pit with a couple of them:

In each case it was the same. At three in the morning there was a knock at the door and two policeman hauled them out of bed and took them off to the clink. There they were subjected to what you'd call treatment until they agreed to change their minds. If they'd gone through the proper machinery, prison it would have been. I supposed they were saved from themselves.

There was overwhelming sympathy for those caught up in the Bevin Boy scheme, some people believing that dispatching the Boys – little more than boys – into the harsh environment of the coalmines was morally repugnant; nevertheless the majority felt that, in war, they should do their duty, and that was how the newspapers saw it. The *Birmingham Post* of 14 March 1944 reported on two cases, one in Sheffield (imprisoned three months), the other in Birmingham (adjourned pending medical examination) and editorialised:

> In each case the objection to work in a pit was based by the defendant on the same reasons; there was no plea of conscientious objection to war work, but a preference for a different form of service. One must feel some sympathy . . . But in each case an individual was demanding that his own wishes should override a direction of the Ministry of Labour . . . If the objector is so self-centred that he can see only his personal point of view, as sharp a sentence as in the Sheffield case may be needed to widen his vision.

Yet there was considerable ambivalence in attitudes, nowhere better exemplified than on the magistrates' benches. Some handed down maximum three-month sentences and heavy fines; others sent men to jail for as little as two weeks – and some imposed fines that were nominal and sent those who appeared before them into the arms of the military, making Bevin retort that 'while I am always pleased to see justice tempered with mercy', the magistrates should enforce the law. 'Courts,' he added, 'are not a place to express their personal view on policy.'

It's doubtful that any Bevin Boy who went to prison served a second term. Having done their time, some decided of their own accord to go down the pits after all. Others continued to refuse – and were quietly shipped off to the army. There was a limit to how tough a government, never entirely comfortable with what it had done, was prepared to be.

Two

Problems, Problems

In early January 1944, days ahead of the first conscripts reporting to be trained as miners, the newspapers were full of encouragement. *The Times* correspondent went to Pontefract in the country's biggest coalfield, Yorkshire, where at the Prince of Wales colliery batches of ex-servicemen and volunteers had been arriving since Christmas.

'"Bevin's boys"', as they are already being called in the mining areas,' he wrote, were assured of 'a cheery welcome'. Looking round, he admired the keenness of the men already there and enthused about the 'first-class pithead baths and splendid canteen':

> Up from a morning below ground, a group of the newcomers noisily washed off the grime under hot and cold showers, and then crossed the road to the canteen to enjoy a chop, roast potatoes, and carrots, and have an outsize mug of tea for 1s. 1d. Having been allowed to share this meal, I can testify that it was excellent.

The conscripts' first visit below ground, he informed his readers, 'will be a thrill'.

On 17 January, when the first 500 conscripts were due to arrive, roughly split between five training collieries – a second in Yorkshire, near Doncaster; one in Lancashire, near Manchester; one near Durham in County Durham; and one near Nuneaton in the Midlands – there was high expectation. Women officers from the Ministry of Labour Welfare Department, wearing red and white armbands inscribed 'Reception officer', bustled helpfully about. At Nuneaton, Movietone News rounded up a bunch of self-conscious arrivals and, once they had collected their pit boots, safety helmets and overalls (the overalls to be handed back at the end of training), blacked up their faces and got them to pretend they were hard at work. Reporters were on hand to ask the newcomers how they felt. At Pontefract the man from *The Times* 'talked with scores of the boys and did not find one who without qualification was pleased to be here'. However, he noted they 'spontaneously added that, now they were here, they would do their best'.

Two days later 'Bevin's boys' were on strike.

The Ministry of Labour had had only two months to set up the ballotee scheme, to reach agreement with the Mineworkers' Federation and identify at which collieries the training centres were to be set up – most at collieries still in production where a pit area could be segregated for training without interfering with the business of winning coal, but a few at collieries worked out earlier in the war and closed down. The Ministry had also had to identify the men to do the training, sort out what the training should be and decide where, once trained, the conscript miners were to go. The newspapers' buoyant despatches gave the impression that all was efficiency and preparedness. It was far from the truth. Six other training centres that should have been open were weeks from completion – leaving Scotland and Wales as yet without the one apiece they were supposed to have.[1] Already hundreds of Scottish and Welsh ballotees as well as others from elsewhere had had their instructions postponed. Even the five centres in operation were hardly ready: workmen were still hammering in the Manchester colliery canteen and the kitchen was devoid of equipment; the trainees had to go to Manchester corporation's British Restaurant[2] for their midday meals.

All the centres were taking recruits by the spring but they were scrambled into service and generally ill-prepared – as late as October, when George Ralston went to the Scottish centre near Dunfermline, he thought it hadn't long been open because he and others were taken away from training and 'spent a lot of time laying kerbstones and making roads'. All the centres were supposed to have pithead baths but they weren't in evidence at the one near Sheffield when Stan Payne, a Cockney saw sharpener in a mill providing timber for ammunition boxes and air-raid shelter bunks, got there in March. As his lodgings had no bathroom and he didn't fancy 'washing the dangly bits behind the kitchen door', he and a couple of other ballotees in the same predicament went into the city on Saturday mornings and indulged in a tub at the municipal slipper baths, although they could barely raise the shilling entrance and the fares. Pithead baths were being installed at Creswell centre near Chesterfield in Derbyshire in March, but still weren't in use.

Creswell, one of the largest centres dealing at any one time with over 500 trainees in various stages of training, was in such a state of chaos that thirty-four ballotees wrote to the Ministry of Fuel (which ran the centres and had responsibility for their employment) protesting there was no lecture room and, a week into their training, they were without helmets and boots. When the situation became public the colliery manager was ordered to hire a local room and the Ministry apologised about the boots and helmets: at this stage around 7,500 trainees were at or had been at the different centres and the manufacturers of both items hadn't been able to keep up with demand.

The trainees had more pressing problems than the readiness of the training collieries: where they were going to live and what they were going to be paid, problems that were very much related.

The Ministry was in the process of converting old army transit camps into hostels and had plans for larger, purpose-built ones, not only to house men while in training but to accommodate them when working in the surrounding collieries, but at the present had managed to open only two. To some extent the problem was alleviated by telling men whose home towns were in public transport distance of training collieries to travel daily. But finding lodgings for the rest put billeting officers into despair; in the already densely populated conurbations of the North and Midlands, most of what was available was long since taken by the thousands of workers directed to the war factories. In Doncaster a search for anyone willing to take 'Bevin's boys' yielded 300 acceptances from 5,000 inquiries. In the Midlands, digs were so scarce the billeting officer had to resort to the Salvation Army. Peter Rainbow, an office boy in a cattle-cake manufacturer's until call-up, came from the village of Yatton near Bristol, alighted in Coventry on his way to Nuneaton 8 miles away, and found himself sent to the Coventry Sally Army dosshouse – the maternity wing of the old hospital next door to the new. So on another occasion was Les Wilson, who had come from Brighton and his job at Sainsbury's. Neither found it an edifying experience.

'God it was a dismal building, and unheated,' says Rainbow.

> The Sally captain was unpleasantly strict in his dealings with the inmates – no gruel if you arrived for breakfast after 6.30am and lights out by 9.30pm. While I was there a battalion of American soldiers arrived in Coventry and about 40 were on the floors above us. On the second night we were awakened by fire sirens and then water poured through the ceiling. The Yanks had had enough of their miserable accommodation and chopped up a few chairs to make a fire, which had got out of control.

'My family was poor but I'd never seen cockroaches,' says Wilson. 'Here, they infested the place. The area we ate in was swarming. When the lights went on they scampered everywhere. I hated them. In the end we collected our breakfast and ate in our room.' He remembers something worse than damp beds:

> All of us in our room got scabies. The authorities hadn't made any health arrangement for us. Undeterred, we went to the hospital next door and they looked after us. We stripped off and got into a bath of some coloured solution. Someone came in and scrubbed our backs – not a pretty nurse, a great big fellow. Our bedding had to be burnt.[3]

Most men were prepared to put up with discomfort in the short term, in the hope that things would get better once they had gone to their permanent employment. But almost all quickly discovered that their wages would barely cover their outgoings, and in some cases wouldn't. Eighteen-year-olds were on 44s a week.

Lodgings typically cost 25 or 30s – but in the Manchester area 35 – with laundry 2s extra. Fares to and from the training collieries were 5s a week on average, with the cost of six midday meals in the colliery canteen 6s 6d. After a stoppage of 1s 10d for insurance, outgoings would come to between 40s 3d and 50s 3d. Those over eighteen were better off on a rising scale of pay that took all recruits to the full miners' wage at twenty-one. But the bulk of the intake were eighteen. And while a one-off settling-in allowance of 24s 6d helped, a lot were broke – and the first week's wage wasn't to be paid until the following week.

In Nuneaton the trainees protested on their second day and were quickly followed by those in Manchester and Pontefract. In Doncaster, they went further, going on all-out strike. Morry Pearce remembers the meeting in the Miners' Welfare gym where he voiced doubt that, as conscripts, they were allowed to strike.

'They want to pay us 44 shillings and our board is 30 shillings, so we're on strike,' one of the lads said. 'Are you prepared to do something about it?' I said, 'I'm on strike with you.' The pit manager was incensed when a deputation of us went to his office and told him we were on strike. 'You can't do that, he said. 'Well look out the bloody window', someone told him. He did – the rest of the lads were sitting in the snow.

The manager did prevent a march to Doncaster: he locked the colliery gates.

In the rush to get the conscript miners into the system, the Ministry of Labour had been very short-sighted. It had sent them in at the rates that pertained for their age in the coal industry, but had overlooked that while young regular miners lived at home, the conscripts had to pay for their keep.[4] The petitions from the four training collieries demanding either a substance allowance[5] or a pay increase that now landed on Bevin's desk made the oversight all too apparent.

Bevin was a man who said and did what he believed was right and he wasn't a man to procrastinate. Four days after the trainees took action he gave them a 16s rise, with another 10s to come when they went underground. The trainees were certain they had won 'because we'd created merry hell' as Les Wilson puts it. Anxious that that wasn't seen to be the case any more than he'd been shamed into action by a *Daily Mirror* campaign, Bevin distanced himself from the situation. The announcement of the award came not from him but in a statement from Lord Porter, chairman of the National Reference Tribunal of the Coalmining Industry. Since early January, the tribunal had been considering the miners' latest pay claim and a statement was imminent. But the timing of its publication on 23 January was no coincidence. The headline recommendation was that the miners should have their third increase of the war, giving them a minimum of £5 underground – and a hastily incorporated paragraph effectively settled the conscripts' demands, though this wasn't admitted directly:

Up to the age of 17½ the wages of boys are increased in the same proportion as that of the men, but in respect of the older youths the Tribunal are of opinion that the change of circumstances is greater and necessitates a larger increase, particularly in view of the fact that as a result of the present compulsory recruitment many of these youth will be working far from home and be compelled to find billets.

'Bevin's boys' got on with their training. Only they weren't 'Bevin's boys' any longer. Their agitation had conferred a kind of status. In the popular newspapers they'd become Bevin Boys – a second capital B and no possessive 's'. The somehow patronising quotation marks were gone, too.

*　　*　　*

Getting the Bevin Boys who travelled long distances to the training centres was another important issue that in its haste the Ministry hadn't thought out. The very first were simply given their instructions and expected to make their way individually, under their own steam. For lads who lived relatively near their designated colliery that wasn't a problem. The rest were less fortunate. Wartime rail travel was tortuous. Many conscripts arrived hours late, after dark. In Manchester they were trickling in up to midnight. Without transport to take them to their lodgings, the reception centre rustled up tea, bread and tinned soup to feed them but, failing to obtain camp beds and blankets, had to put them up at the town hall with the firewatchers. Altogether, as many as one man in ten failed to arrive on the first day.

The Ministry learnt from yet another mistake and got long-distance conscripts to travel in batches when numbers made it possible. That helped. So did putting reception officers on the platform at the three London termini, St Pancras, King's Cross and Euston, from which Bevin Boys went north to coal country, as well as on the main destination stations. But rail travel remained a nightmare and in the first half of 1944 was, in fact, the worst of the war: the lines were clogged with military traffic in the great build-up to D-Day. Passenger trains were covering a third less mileage than before the war but squeezing in more than twice as many people, most of them troops who slept in the corridors and even the luggage racks, their kit everywhere. Many coaches didn't even have corridors and therefore were without toilets. It was not without irony that the poor quality of much of the coal reduced the performance of locomotives, which broke down frequently. After dark, train travel was a disorienting experience: all but the smallest stations had had their name boards taken down in 1940 and it was almost impossible to know where you were. 'If you can't see the name and can't hear the porter's voice – ask another traveller,' one railway poster advised. 'If you know where you are by local signs and sounds – please tell others in the carriage.' Right to the end of the ballot, some conscripts continued to turn up hours late, stumbling off trains unsure whether they were in the right place or not.

For many Bevin Boys the very business of travelling brought home their plight: all around them contingents of servicemen jostled and clamoured – a painful reminder that theirs was to be a different kind of war. David Reekie looked enviously at the uniforms that came and went around him as he stood on the concourse at King's Cross with a straggle of other Bevin Boys on April Fools Day,

> and what a bloody fool I felt. Our cases and persons were adorned with large tie-on tags that were over-printed with the Union Jack. Four years of war had taken toll of the fabric of the station and little or no glass remained in the roof. Smoke and soot had stamped its presence on every exposed surface. It was in some way appropriate for us fledgling coal miners – we were condemned to a similar environment underground.

Some had very personal reasons for reflecting on the unfairness of life. Alan Gregory might have taken up the Exhibition in classics he had won to Cambridge, but had decided to get into the war, applying for naval 'Y' course entry, which would have meant six months combined study and service at a university, followed by a commission. But he applied too late, 'missed the bus and had to await normal call-up'. He left home in Forest Hill with the 'rotten lucks' of his fellow players in the Dulwich College rugby team – and their laughter. There were other men who had hoped for a service commission and a possible career. Bill Gibbs, a bank clerk from Grimsby, was one. He had tried to enlist at seventeen and a half but his mother refused to sign his papers. 'So I made my sorry way to Creswell colliery: train to Worksop, East Midlands bus the remainder. At least my mother was elated.' Derrick Warren, who worked at Battersea power station as a boiler house control assistant, was another. He had passed the entrance exam for officers' training college but, like Alan Gregory, missed a cut-off date:

> The intake was up to 31 July – my birthday was 8 August. So I had to wait for the next. In the meantime Ernie came and got me. A pal of mine from Wandsworth who passed with me went in and was commissioned in REME as I wanted to be. Bloody annoying.

Ronald Garratt, deprived by the ballot of a navy posting, on his way from Birmingham to the Nuneaton training centre via Coventry, had the misfortune to catch a train crammed with hundreds of sailors. With two other Bevin Boys he found a seat in one of the compartments.

> 'Hello, lads, on the way to Pompey?' they greeted us when we got in. They were shocked to hear we were being sent to work down the pits. To cap it

all one of them had a copy of the *Daily Herald* carrying a full front page on mining under the headline MINING IS HELL. Just what we needed to cheer us up. We pulled into Coventry all too quickly for me. 'Stay on the train and we'll sort it out in Pompey,' the navy kept saying and they were very persistent. I was very tempted.

John Squibb, who'd been working in his father's small printing business in Weymouth, boarded the late-night train to Waterloo on his way to training in Durham and was discomforted by the uniforms around him, but these were worn by a group of GIs who were accompanied by their girlfriends. 'The subdued lighting on account of the blackout suited their amorous intention very well but deeply embarrassed me and depressed me more than I already was. No glamorous uniform for me. I had a suitcase of old clothes I'd been instructed to bring to wear down the pit.' So depressed was Stuart Chislett from Epsom, a clerk in a rubber company, when his number came up in June 1944 that 'odd as it may seem now, I had no thought that this was the moment when the D-Day invasion of France was taking place'.

Journeys that might have taken three or four hours often took ten or twelve. Norman Brickell, just finished his printing apprenticeship in the family business in Gillingham, north Dorset, travelled on his own to Pontefract, where he was due at 8.30p.m. Bevin Boys who travelled in groups from London got a reserved carriage; those who made their own way from elsewhere in the country as often as not had to stand. Brickell did, all the way, eventually being

dumped on the station at 2.30 in the morning. Completely blacked out. Not a light to be seen. I headed off over the echoing cobbled street to the police station and told the sergeant what I was supposed to be doing. 'Tha' better come in here then, lad, I've got a bed for thee,' he said. So I spent my first night in Yorkshire in a police cell.

Eric Ward, a Southampton insurance clerk, spent his first night in County Durham in yet another Sally Army.

With my usual luck I got to Newcastle at night and didn't find any representative there to meet me in accordance with the instructions. I re-read these and found I should report to the RTO [railway transport officer] if I was in trouble. He advised me to try the merchant navy hostel and travel to Stanley in the morning. Where on earth was the merchant navy hostel? Outside the station a kindly pedestrian gave me clear Geordie directions. I rang the bell. The door opened, I got a curt 'Sorry, full up', and the door closed. Back to the RTO. Try the Salvation Army. I got a bed there, in a dormitory with partitioned bed spaces, where the snoring was definitely not in musical unison. And with my usual luck the hostel was close to a large church where the clock chimed the quarters and the hours all night long. By six in the morning I'd had enough. I crept out to find somewhere for a cup of tea and sandwich and then the bus station.

Ward was inclined to think his luck had run out a few weeks earlier. He'd taken a day's holiday and gone on the train to Bournemouth with his mother. Coming back, they shared a compartment with one other occupant,

> a large-built gentleman busily working through a sheaf of papers on his briefcase. My mother was a person who, despite a pronounced stammer, could always start a conversation with a stranger. Within 15 minutes, with her customary directness, she said: 'You're Ernest Bevin, aren't you?' He admitted he was. The rest of the journey was spent on general talk. As we approached Southampton Central my mother made her parting remark: 'Very pleased to have met you, Mr Bevin. This lad has been passed A1 for the army. Don't you dare make him one of your Bevin Boys.'

Even arriving in daylight didn't necessarily solve trainees' problems in reporting. Hard-pressed Ministry of Labour staff were often busy elsewhere and not at the stations; the reception centres, if the trainees managed to find their way there, didn't always have transport. In either instance arrivees had to make their own way. 'When I got to Doncaster a good number of Bevin Boys got off the train with me and we hung about wondering what to do,' says Les Thomas.

> We had to ask local people how to get to Askern colliery, which was some miles out of Doncaster. It was just left to us to queue up at the bus stop with the housewives with their shopping, and pay our own fare. This was a bad start and a shambles on the part of the Ministry. We'd been travelling since early that day without anything to eat or drink. We didn't get a meal until the evening – and we had to pay for that.

By July 1944, forty-four hostels were open, with a capacity of 15,000, and the majority of trainees could be accommodated in them. But in the first months, intakes had no choice but to go into digs. At least a few Bevin Boys struck lucky: some communities, particularly in the villages and small towns, rallied round. Gerald Carey, a clerk in a cold storage company's estates department based in the City of London, who was on his way to the Creswell centre, was met at Chesterfield by 'a large man in plus-fours', who handed him and his fellows over 'to several portly ladies of the Women's Voluntary Service', who took them to their village hall in Woodhouse, Mansfield, where they'd prepared tea and biscuits – an extraordinarily nice gesture. They checked our rations books for us, changed the addresses on our identity cards and told us where the billets they'd arranged for us were. After we'd finished our tea the vicar gave us a little talk on how we should behave – I think he was suspicious of our morals though I can't think why.

In Mansfield itself, where he arrived as part of a different intake, John Potts, a post office clerk from Wellingborough in Northamptonshire, found himself in 'a home from home',

> above a small corner shop run by a mother and daughter, the pair of them little round dumplings, who prepared wonderful cooked breakfasts at 6.30 every morning, and for tea would get stuff out of the shop, like cream cakes that I hadn't seen for years. Sunday lunches were out of this world. The mother's speciality was onion gravy, which I'd never seen before.

For Desmond Edwards, his lodgings in Blackwood near Oakdale training centre in Monmouthshire were a revelation: 'Having been brought up in a terraced house, one of a very long row, in a miners' valley, a modern house in a residential area high up on the plateau with trees along the curving roads was gracious living.'

Before the Bevin Boys began going to the coalfields, the Ministry of Labour had rather grandly announced that it 'proposed to grade billets and fit men into an appropriate kind of household'. When it came to it, the scarcity of accommodation was so great – and greater still with each intake – that billeting officers had to shut their eyes and take what they could get; and sometimes what they could get was deplorable. Fleas were a common experience; many houses had as many cockroaches as the Coventry Sally; and perhaps most had no indoor sanitation. 'The couple I stayed with when I was training in Wales were very nice – he was a guard on the railway,' says Jack Garland, who'd worked in the Bristol office of a company making cardboard for cigarette packets.

> But their little terrace was primitive. The WC was out the back and there was no flush. You had to carry a bowl of water with you. The house did have electricity, which a lot didn't. But she'd only put a shilling in the meter when the lights went out. I used to go to bed with a candle. If the electricity was on you clicked the switch in the kitchen standing outside the door to give the cockroaches a chance to scuttle under the skirting.

One landlady in Castleford, with a family of ten living in a three-bedroomed house, took in four Bevin Boys. Elsewhere, another packed six into one room. Geoff Baker found himself one of nine taken into a house in Stoke, and the only one without a bed.

My planned accommodation for some reason wasn't now available and I remained miserably on the coach as it dropped off others with various landladies around the Potteries. At the last port of call this lady had agreed to take eight and really didn't have room for me, but she took pity and let me sleep on the settee. I then struck a deal with a night-shift worker in the house to share his bed – we left it warm for each other. I do remember thinking 'I'd better not let mum know about this.'

At least the pair didn't have to share the bed at the same time. Many trainees did, usually with a fellow trainee, but sometimes with someone else.

Conditions were a shock to the system for those from middle-class and respectable working-class homes, none more so than for David Reekie. Met at Leeds Central by a reception committee with various means of transport at their disposal, he was thrilled to be one of four Bevin Boys driven in an old Rolls Royce to the mining village where they were to be billeted. He was less thrilled when he was deposited on the pavement outside a council house in Skellow where, he thought, 'one of His Majesty's houses of detention would have blended in well'. And even less thrilled when he met the family he was to stay with:

> Truly, truly a nightmare family. He was small, a surface worker at a local colliery, and she was big, with a plaster that was in a disgusting state on her right arm, with which she frequently hit him over the head. They could never have washed, the smell was terrible. I was honestly terrified. I can't tell you how glad I was to get out of there.

Many who took in Bevin Boys did their best. But some exploited their young lodgers outrageously (and even searched their suitcases for any foodstuffs worth taking). Trainees were supposed to get a cooked breakfast, high tea on returning from the training colliery, a hot drink before bed, and a two-course cooked dinner on Sundays. Many got meagre fare, watered milk, stale bread and tea from leaves that were dried out time and again. Many had to spend the precious few shillings in their pocket on extra food.

Some Bevin Boys made official complaints and billeting officers probably struck off those landladies who were the worst offenders. But in four weeks the trainees moved on – and the headache of finding accommodation for the next intake started again.

*　　*　　*

When was a Bevin Boy a Bevin Boy is something of a conundrum. The Bevin scheme embraced ex-servicemen and conscientious objectors, optants and volunteers, and ballotees – who considered themselves the only true Bevin Boys.

The ballotees' view of optants as men who were avoiding the fight that they had so much wanted to be part of, was largely unfair.[6] Some optants undoubtedly were anxious to avoid combat; others chose mining in preference to the forces because they lived in or near a coalmining area and could live at home or get there at weekends. But many, of an age with the ballotees, had answered Bevin's call to go down the pits because they believed it was a better way to help their country. (And as it became apparent in early 1945 that the war would soon be over, increasing numbers of conscripts opted for the pits precisely because there wouldn't be any fighting to be part of.) Tony Brown, a signwriter from Leicester, felt like that, though his position was complicated by being a communist at the time; 'holding certain views with the crass certainty of immaturity', he emphatically did not want to be part of the military. 'But I did think I had a part to play: I thought I could do something for the people of this country as a miner – the country was crying out for coal.'

It could be said that Michael Edmonds became an optant on a whim, and John Wiffen because he lost his temper.

A trainee architect from Bere Regis near Wareham in Dorset, Edmonds had spent a year in Bristol, attending the Royal Academy of Architecture, and his landlord, an ex-commercial traveller whose patch had been south Wales, had regaled him with tales of miners and mining. Edmonds had found these 'upbeat and intriguing'. But what sold him on the idea of going down the pits was his interest in geology: 'An only child in a one-parent family, I'd wandered around the Dorset countryside, visiting stone quarries, pottering about the chalk bays on Purbeck, collecting fossils and bits of rock. When it came to conscription I thought: surely mining isn't so far from geology . . .' Wiffen, from Meols on the north Wirral coast, a year done at Liverpool University, went to his registration intent on getting into the navy – but on his terms. Some of his friends had gone in ahead of him and 'spent month after month on the parade ground or saluting the flag or kicking their heels in a foreign port on the equator, waiting for the word to steam that never came'. Determined to avoid a 'bullshit destroyer' or 'getting cooped up in a naval barracks', he asked to be posted to a frigate, an MTB or a minesweeper. The naval officer he faced might have said, 'We'll see what we can do.' Instead he told Wiffen: 'We can't have you young chaps picking and choosing: you'll go where you're sent.' To which Wiffen responded: 'In that case, perhaps you could direct me to the office responsible for recruitment to the coalmines.'

The majority of volunteers were in their thirties and forties, mostly men who disliked the war work into which they'd been directed and thought mining was preferable, or those who'd been directed to other parts of the country and were unhappy far from home. More than a few shopkeepers whose premises had been destroyed in the bombing also became volunteers and, perhaps a very, very few lads over seventeen but not yet summoned to registration who answered Bevin's call (though as the MP Seymour Cocks said in a Commons' discussion on the decline of young boys going into the pits full time, 'no one would today unless he was ultra-patriotic or half-witted'). Uniquely, however, some volunteers were men of the ballotees' age – German and Austrian Jews who as children in 1939 had escaped to England from Nazi persecution.[7]

One was Victor Simons (then Schneider), who'd arrived from Berlin, been taken in by a family in Wargrave, Berkshire, and gone to Shrewsbury for his education. At seventeen, 'as an enemy alien obliged to do work of vital national importance', he was a labourer on a farm in Hemel Hempstead in Hertfordshire, but wanted to fight. 'I owed a debt to this country,' he says. He tried and failed to get into the army and the navy – 'a temporary embargo on recruitment in the army at the time I applied, rather oddly; the navy would not take me as I was not born of British parents'. As poor eyesight would have disbarred him from flying, the RAF wasn't an option as far as he was concerned. So he chose to become a miner.

> I could have stayed on the farm, but I chose not to. Labouring, I was not contributing enough to the war effort and I saw my contribution as a responsibility. There was also, frankly, another reason. I was without capital – my father was a lawyer in Berlin but with my mother was now in domestic service in England after a year in internment. Thrown out of one country I needed to find a profession in another. And I thought that, if I became a miner, with just a little grey matter, I could go to technical college and get a colliery manager's certificate.
>
> I am always very particular in referring to myself as a coalmining volunteer – I was not called up. I do use 'Bevin Boy' in explanation of my war service.

Meir Weiss, who'd escaped from Vienna, was another volunteer and how he became one is a story with an unusual twist: he went down the pits as a member of the only mining kibbutz in Jewish history.

Son of a travelling salesman, Weiss like Simons went to work on a farm (in his case between Ashford and Maidstone) as a member of a small group of young Jewish refugees. There were numerous small communities like it called *Hachsharot*, meaning 'preparatory': those who lived in them were preparing for life in the kibbutzim when they could go to Palestine and the new Jewish state. Mostly the *Hachsharot* were housed in rundown country houses – there was no shortage of such places during the war – and did agricultural work, frequently moving on. Weiss laboured on farms not only in Kent but in Devon and Shropshire, 'working hard by day, arguing politics and learning Hebrew by night, reading by paraffin lamp and dreaming of building a better world'.

In 1944 he and others with him reached call-up age and ten or so went off to the army – this during the period that another move took the community to a farm at Stalybridge near Manchester. The group as a whole now faced a dilemma: 'We wanted to stay together, not to be dispersed,' says Weiss. 'There were 25 or 26 of us, I think, half boys, half girls, and we wanted to go to Palestine together when the war was over. If the group broke up entirely we would not have been able to achieve that.'

With their registration imminent, the young men in the *Hachsharot* came up with a bold possibility: ask the Ministry of Labour whether, if they volunteered to go down the pits collectively, they could continue to live in a kibbutz. Surprisingly the Ministry agreed, and got the Ministry of Fuel to stump up the money to knock together and furnish two large Victorian semis between Stalybridge and Ashton-under-Lyme. The premises were consecrated in August. And Weiss and his comrades went to the nearby training centre and then to work at Ashton Moss colliery, a few miles away between Stalybridge and Ashton. Weiss looked forward to becoming a miner, explaining laconically: 'I had read Emile Zola's *Germinal* and seen the film *How Green Was My Valley*.'

Ballotees and younger optants tended to assume that those perhaps half a dozen years older than themselves were all ex-HMF. Most were. But some were men whose job had been de-reserved, like Alan Brailsford, who at twenty-five was in the oldest age group to be called up under the ballot scheme. An accountant at a Sheffield steelworks, he'd been responsible for the capital administration of a bomb factory his company was ordered to build on a greenfield site near Rotherham. At the end of 1943 the factory was closed down: the bombs it produced were no longer big enough. Brailsford was convinced he was in the navy; the ballot said otherwise, 'rather poor recompense for having worked every hour possible seven days a week, getting in 5,000-ton presses, sorting out the production line – and as a Home Guard, the only NCO who hadn't been an old sweat in the First World War, even guarding my own factory'.

The reason why most men agreed to come out of the army for the mines was because, after three or four years' active service, they'd had enough. Others, of course, had personal reasons for wanting to be near home. Some were misfits that the army wanted shot of and 'volunteered'. Most of those who left the air force had been thwarted in their desire of becoming aircrew through failure, colour-blindness or injury, and couldn't face the idea of being 'downgraded' to some utilitarian role or being transferred into the army. Chronic sinusitis, for which he had a double nasal anstrostomy between being selected for aircrew and enlistment got the rep actor Brian Rix downgraded and sent to the RAF re-employment centre on the Isle of Sheppey, where he was offered cook, medical orderly or transfer into the pits. He chose the pits, arriving at Doncaster training colliery 'dressed in my lovely grey chalk-striped suit and green porkpie hat which clearly showed I had been in the forces'.[8]

How many conscientious objectors went underground is impossible to ascertain – numbers are hidden within general totals. In late 1944 the Home Secretary,

Herbert Morrison, told the House that the figure was forty-one – which raised questions in the newspapers as to why more COs (60,000 in the course of the war) weren't directed into mining.[9] It seems probable that Morrison's figure was only of those like Jim Bates, a Londoner who'd just left school, who was given the option of the pits on call-up: just ahead of the ballot scheme, Bevin ruled that men registered by a tribunal for non-combatant duties in the forces could choose mining instead, which Bates did. But scores of those who left the army for the pits were conscientious objectors serving in non-combatant units (though a number were in the Medical Corp attached to the Paras!). It's at least a possibility that a conscripted Bevin Boy or two were conscientious objectors: when he instituted the ballot, Bevin additionally ruled that COs who failed to get conditional exemption at tribunal, as well as those who were registered for non-combatant duties, would in future go into the hat.

There is considerable anecdotal evidence from Bevin Boys of conscientious objectors among them; Geoff Baker, for instance, says that of the eight other trainees in his training digs, seven were COs. There appears to have been a mild hostility between some conscript and some optant miners in the early stages of being thrown together, but conscientious objectors were generally accepted and their views respected by most Bevin Boys – if not by Brian Evans. After training he was posted to a pit not far from Stourbridge and found:

> there were very few Bevin Boys, real ones I mean. I say real ones deliberately because there were quite a few there who were 'conchies': young men who'd chosen to go into the pits rather than do service in the expected way. Couldn't stand them I'm afraid – in fact I had a confrontation with one. I can't remember exactly, but something he said or did riled me. He was a Londoner for a start. At that time people outside London disliked Londoners, it was inborn. We considered them cocky and too smart for their own good. And I'd heard him call himself a Bevin Boy, which as far as I was concerned he wasn't, he was a conscientious objector. Anyway, at pit bottom at the end of a shift, something . . . I went for him, cheered on by the watching miners waiting for the cage to go up top. It wasn't kicking and gauging but not quite handbags at dawn. There were a few strong punches.

It's hard to say whether the most extraordinary man to become a miner in the Second World War was an ex-serviceman or a volunteer: paradoxically, John Platts-Mills was both.

A New Zealander, Platts-Mills, a barrister with two first-class honours degrees (in law from Victoria University College, in jurisprudence from Oxford), had married well, was well-connected and had a thriving career. Recommended to the RAF by his country's High Commissioner, he was commissioned soon after the outbreak of war but failed to get a posting. Probably the security services kept him away from active service because of his radical political views; he, however, was certain that he was 'excluded from any form of normal war service by the stupidities of Bevin' (the two had clashed) and the narrow-mindedness of Sir Archibald Sinclair, the Minister for Air, who 'thought it quite treasonable that anybody on the left should dare infiltrate himself into the RAF under his command.'[10] A serious shortage of men on vessels crossing the Atlantic now prompted Platts-Mills to volunteer for the navy and he was chosen for the officers' training school at Dartmouth, but his orders failed to show up, 'so Ernie Bevin's technique which had kept me out of the air force acted within days to protect the navy from me.'

When Hitler invaded the USSR in 1941, Churchill, whom Platts-Mills knew socially, sent for him. Since 1918, he told Platts-Mills, he'd been teaching the British people that 'the Russians are monsters . . . They eat their own children . . . Now I've got to change all that.' Or, rather, he wanted Platts-Mills to do it. For the next two and a half years Platts-Mills and the team he gathered together ran around the country doing such a good job that 'almost every big gun produced had written on it in red letters GUNS FOR JOE. And tanks were labelled TANKS FOR JOE'. Indeed the team did such a good job that Churchill's nose was put out of joint. Summoning Platts-Mills to his room at the House he said: 'Look at these pictures in the newspaper. This is what you've done. "Tanks for Joe". Why not tanks for me? Why not "Tanks for Winnie"? You've gone too far. You had better go and do something else. What do you want to do?' In reply, Platts-Mills told Churchill that 'if he was fed up with me I'd go down the mines. He said, "Well, you'll never get another chance, so you'd best go and do that if you want to," then slyly added, "Ernie can't chuck you out of that."'

Platts-Mills' military enthusiasm had been blunted by his experience with the RAF and the navy, but he now received a tentative approach from the army which, eventually, he cautiously accepted. He was instructed 'to report to a named unit at Colchester', his travel warrant and subsistence allowance enclosed. Within days his instructions were rescinded – perhaps Bevin had struck again.

In July 1944, Platts-Mills reported to Askern Main colliery in Yorkshire. He was thirty-seven.

Ballotees, optants, volunteers, servicemen – the public were not concerned with the niceties of differences between them. Much to the chagrin of the conscripts, they were all loosely regarded as Bevin Boys.[11] In 1944, according to the records, in addition to the 15,363 conscripts, 8,668 optants, 4,745 servicemen, and 3,139 volunteers (and three of Morrison's conscientious objectors) went into the pits with them. In 1945, up to the termination of the ballot in April, 5,534 conscripts, 6,989 optants, 1,906 servicemen, and 1,475 volunteers (and the rest of Morrison's forty-one conscientious objectors) joined them.

*   *   *

What training recruits should get was a vexed question because conditions and customs varied so greatly in the mining industry, not just from coalfield to coalfield but from pit to pit. In the main these were attributable to geology and what the colliery owners had invested – or not. Seams could be 40ft high or as little as 12 or 14in, some virtually flat, others snaking up and down through the earth – the tunnels with them. Almost all pits were entered by a shaft with its alternating pair of cages, but in some (called drift mines) men walked down a slope into them. Some pits employed 3,000 or 4,000 men; others fewer than 100. Some were shallow as, for example, the majority in County Durham; or deep, as in Yorkshire and Kent. But where the Yorkshire pits were dry and moderately warm, Kent's, extending under the Channel, were wet and fiercely humid.

Up to the war, mechanisation was being gradually introduced and in varying degrees was in two-thirds of collieries, but had been halted by it: British manufacturers of mining machinery were turned over to munitions. The contrast between pits still working the nineteenth-century method of 'pillar and stall' and those that worked the 'longwall' method was startling. In the former, the coal was, as miners said, 'hand got' – dug out with pick and shovel, the colliers burrowing into the seam, leaving giant blocks beside and behind them to support the roof. Unless a pit roof was notoriously unstable, shot firing loosened the face. In the longwall method, all the coal was extracted, mechanical cutters slicing into the bottom half of the face along a 100yds or more to a depth of 2 or 3yds, the colliers, again assisted by shot firing, then bringing down the rest. Longwall pits had a variety of mechanical shaker pans, jiggers and conveyor belts to load the tubs, which ran on small-gauge rail tracks to the pit bottom, attached in sets to a moving steel cable powered by a stationary engine or a series, depending on the distance from the coalface, in most cases driven by compressed air. A handful of the most modern pits had diesel locos instead of stationary engines – but at the other end of the scale, a number still used ponies (or cob horses in the relatively small number of pits with high seams and therefore roads) and in a few, men did the hauling. Yet other pits operated a mixed system, ponies hauling tubs to and from marshalling points from which cable haulage took over the rest of the way. Tub size depended on the clearance room of the roadways – which in turn depended on the height of the coal seams – and carried from around eight hundredweight to as much as a ton and a half.

By the time they were eighteen and of an age with most Bevin Boys, lads who'd followed their fathers and older brothers into the pits had four years' experience. How much training did Bevin Boys need? Speaking in the House in December 1943, Lieutenant Colonel C.G. Lancaster (Fylde, Unionist, and a colliery director) supposed it would be possible, by shortening the stages of training, to get 50 per cent of Bevin Boys in full production at the end of twelve months, the other 50 per cent after eighteen.

The Ministry of Labour, after consultation with the Ministry of Fuel, announced a training programme of six weeks: four at the training centres, to be followed by a two-week induction at the pits where the Bevin Boys were sent to work. Half the preliminary four weeks was to be spent on practical work, the other half split between classroom theory and physical toughening up. A total of six weeks was precious little time in which to ready young men for coalmining. Bevin remarked: 'The war will not wait.'

All the miners who became instructors had twenty or more years in their industry and had either been pulled away from their jobs or brought out of retirement. Given some hasty training themselves in how to lecture, they taught the trainees about how a mine was worked, the dangers of gas and how to test for it, the causes of underground fires and, over and over, how to stay safe. And then they taught them to put theory into practice. They had been selected as much for their cheerfulness and patience as much as their know-how. And, as Ian McInnes remembers, they were listened to by most of their 'less than enthusiastic' charges – even if those from outside a particular area and especially those from the south of England sometimes found their Scouse, Geordie, broad Yorkshire and Rhondda Valley English, heavily laced with pit vernacular, impenetrable.

Stan Payne was one of those who arrived with a chip on his shoulder and was won over. Like Anthony Shaffer he was convinced that an outside agency and not a number drawn from Bevin's hat was to blame for his becoming a miner:

I'd chosen the navy as my older brother was a leading seaman and would have claimed me to serve with him, which could be done at that stage. However, my interview was with a Marine sergeant who said it could only be arranged if I signed on. I wasn't going to accept that so the sergeant got a bit huffy. So I told him in no uncertain terms what he could do with his navy and just who the fuck he thought he was talking to. Nothing would convince me that my call to mining wasn't a direct result of this episode.

But at Sheffield he grew interested in what he was going to be doing because of 'our underground instructor, named Albert Shaw, who was exceptionally friendly and helpful – he became a real friend. He lived at Hackenthorpe, a small village about three miles away across the moors, actually in Derbyshire. Two or three of us would walk to his home on Sunday mornings to visit.'

Another who committed himself was Morry Pearce at Doncaster, thanks to an instructor called Albert Chambers,

who condensed a lifetime's experience into a month and managed to knock into our heads what we needed to know to survive in a dangerous environment. He was on loan to the training centre from Bentley colliery where me and others were being sent to work and he was going back after finishing with our intake. Saying cheerio I said I'd seen him there – but he was decapitated by a steel roof support that snapped before we arrived. In many ways he was like a second father. I'll never forget the raw emotion me and the other lads felt. A terrible welcome to our new colliery.

Because the Bevin Boys could go to differing types of pit, the practical training of necessity was general, including everything from fitting pit props to harnessing ponies; but as most were likely to be employed on some form of mechanical haulage, this was the area of concentration.

It was only down the pit, when they found out what was really involved, that the Bevin Boys realised what physically hard work was in store – and how potentially dangerous. After a demonstration of how to lash linked coal tubs to a moving cable with a short length of chain attached to the front and rear of the set, Reg Taylor, a civilian worker with the police in Bradford, was the first in his group asked to see what he could do.

> I tried to pick up one end of the chain. It did not want to leave the ground. I had never trained as a weightlifter. I managed to get it up to shoulder height, desperately trying not to trip up over the rails as the three tubs challenged me with their own forms of dumb insolence. With one mighty effort I managed to hurl the hook over the rope and eventually hooked it on to itself to an accompaniment of guffaws from my fellow trainees. It came as a shock when the instructor told me to run after the tubs and lash the last tub to the rope.[12]

To stop the tubs, the lashing chain had to be released by hitting it in just the right place with an iron bar, and that was the trick – hitting it in just the right place. The instructor showed Tony Brown's group how and when it came to his turn:

> I took a hefty swipe at the chain and missing, hitting the rope [cable] square on. A moving steel rope under tension doesn't like being hit with an iron bar and this one bounced back, smacking me hard over the eye. I saw stars – and me and my helmet went flying.

The nurse at the hostel bandaged Brown's head. 'When the new batch of trainees arrived they looked very worried indeed, which I enjoyed immensely.'

Introduced to a different system in which tubs were clamped to the cable, John Etty, a junior clerk in the social services department of West Riding County Council and a part-time professional rugby player, was in the middle of his turn when the instructor called time to finish.

Everyone went towards the cage. As I'd started my task I completed it. The tubs moved off. But so did I: I'd fastened a finger in the clamp. The overtub rope ran through horizontal wheels on the wall and one was getting nearer and nearer. I didn't know what to do, my mind was blank. Suddenly the rope stopped – the engine had evidently been switched off: the end of training. Someone up there must have liked me. In a few seconds more I'd have had a mangled hand and no future in rugby.

As Bevin Boys were to discover, tub-sets frequently jumped the rails in the pit and heaving a tub filled with coal back on to the track was no joke. It wasn't so much about strength but technique. 'You had to bend your knees with your back against the tub and bodily lift it back on,' Ian McInnes explains.

Local newspapers across the country carried small pieces about the first of the Bevin Boys from their areas. If the phrase 'set off cheerfully', which appeared in most, including *The Surrey Herald's* announcement of the departure of Jim Ribbans, an engineering student from Addlestone, was something of an exaggeration, that the majority of conscripts intended to make a go of their new life wasn't. Indeed, some set off seeing what lay ahead as an adventure. 'It's a far cry from a white apron to the black coal mine,' Les Wilson's father told the *Brighton Evening Argus*, 'but Leslie assures me he will do his best.'

A minority, however, especially in February and March 1944 when the first batches of those who'd failed appeals came through in numbers, had no intention of knuckling under. Most Bevin Boys arrived at the centres expecting some sort of labour battalion organised along army lines. Quickly discovering that wasn't the case – they'd signed on at the local labour exchange, which clearly proved their civilian status – some refused to cooperate. In lectures they declined to take notes; when the lights were turned off for lantern slides or Ministry training films, they went to sleep. On one occasion Roland Garratt started to watch a film 'in a room full of trainees, but when the lights went on at the finish only six of us remained'.

The physical training instructors, all of them ex- or serving army and air force NCOs,[13] had more widespread indiscipline to deal with than the miners because many trainees found their daily exercises silly and irrelevant. The Ministry publicly spoke of 'training specially designed to develop strength and agility and in particular those muscles which the men will need to bring into use in their work below ground'. Geoffrey Mockford at Bolsover found himself 'doing the kind of thing done in junior school at the time: jumping up and down, raising our hands above our head; small games like weaving between other members in the line. I thought it a farce – and from what I've heard it was much the same in other centres.' It was of some tart amusement that the 'special apparatus', which the Ministry claimed would 'help men become more proficient in lifting, carrying and hauling', turned out to be tree trunks and girders that the trainees were supposed to carry about in teams.

Some later intakes of Bevin Boys were issued with gym shoes and shorts for PT, but most, including all the early ones, had only their pit boots, each weighing 2½lbs, and their overalls, which 'had everyone fallingaround in hysterics,' Mockford adds. 'We went running in the same outfit, with roughly the same result. You couldn't run in pit boots. So our progress along footpaths, small roads and muddy fields was a mixture of a slow trot and walking.'

Men who'd enjoyed drill in cadet corps or the Home Guard found they couldn't march in pit boots either; as one Bevin Boy remarked, they made army boots feel like ballet slippers. David Reekie:

> I was used to wearing army issue boots and once marched 20 miles with full pack and a rifle-grenade launcher without so much as a blister. This footwear was something else: huge shiny steel toecaps married to soles and heels of such armour as to make movement more appropriate to novices at a deep-sea diving school. The tremendous, ragged clatter of 50 pairs of heavily studded boots around the local roads had to be heard to be believed. We were a shambles. When we came to a metal road bridge over a railway line we disintegrated into a melee of skipping, giggling idiots.

He remembers '25 a side football. Can you imagine a crack on the shin from a pit boot? There were casualties. I think I was the first person to develop the technique of running off the ball.'

Owen Jones, an apprentice cabinetmaker from Portslade near Brighton and self-confessedly the least military of men, found marching around the streets deeply embarrassing. 'Creswell village was just houses in a circle so you couldn't escape the curtains being drawn aside. Everyone watching: I felt very sensitive about it. And what was the point in learning to march at all? What had marching to do with working down a pit?'

Most men felt exactly the same about the boxing they had to endure in the cause of toughening up. It pleased Stan Payne no end that one of the PTIs got his comeuppance:

> Alan Brown was ex-army and became manager of Preston North End after the war. In the gym one day they paired us off, roughly by weight, for sparring. One of the group was a big gentle fellow named Dick Sherrat. Dick weighed something over 14 stones and had been a warehouseman, quite used to lifting and carrying heavy weights. When it came to his turn for the gloves he made it clear he didn't want to fight and, in any case, there was no one his size to take him on. Mr Brown insisted and pitted himself against Dick. Still Dick was reluctant and put no heart into the proceedings. Mr Brown then lands a punch that hurt. Gentle Dick cuts loose with 14 stones behind a strong right arm.

End of sparring session for our group, Mr Brown sporting livid black eye for several days.

In service life the PTIs had the threat of military punishment – doubling a man around a parade ground with his rifle over his head, handing out a spell of jankers in the cookhouse – to back them up, which was a powerful disincentive to defiant behaviour. In the training centres, however, in a civilian environment, they had only their presence and powers of persuasion to control high-spirited, or wilful, young men; and sometimes these weren't enough. Out on runs or other exercises, some Bevin Boys simply disappeared. Roland Garratt:

> One exercise was carrying steel pylon parts up and down the sides of hills formed from the waste from the pit and we were supposed to assemble at the top of the last. I was in the third team. When we arrived exhausted at the top we were very surprised to find no one there – the previous teams had continued down the other side, over the hedge, and caught a bus to Nuneaton to visit the local cinema.

Les Wilson went on a 5-mile run with others and caught a bus back, but he found a way of getting out of most of his training:

> At the colliery the cook, a woman of ample proportions, complained bitterly about the extra numbers. At this, somebody asked for volunteers and I made sure I was one of those chosen. For the whole month I hardly did anything else but help in the cookhouse during the morning. And we ate in the kitchen. 'You don't eat out there with the others, you eat at the back, here with us,' cook said. Looking at her size I thought this just had to be good. On top of that I didn't have to pay for dinners.

More than a few Bevin Boys simply left the centres and went home, including Frank Pratt. Still angry that he'd been denied compassionate exemption to look after his mother, he discovered when he got to Oakdale training colliery in Wales that he had a different number than all the other conscripts:

> I went round the lot, and all of them were a nine. I was a two. I shouldn't even have been there. I didn't complain about that: I thought I might somehow be labelled a conscientious objector and that was the last thing I wanted. I can only conclude I was where I was because my surname is Pratt.[14] But I wasn't giving in to them. You can take a horse to water but you can't make him drink. Talking to the lads who were already there, received wisdom was, you could take a week off in training – do a couple of

weeks, get away for seven days. It took that long for the police to catch up with you. So I scarpered, as a lot more did.

A handful disappeared altogether, their fate unknown to their fellow trainees.

John Potts, a post office clerk from Wellingborough in Northants, was at Creswell training centre in Derbyshire when the Duke of Gloucester came to inspect it – 'the questionable highlight' of Potts' training. 'The really bolshie Bevin Boys,' he remembers, 'were sent down the pit. I was on top, in the group demonstrating our fitness, slinging pit props around for what seem like hours. The Duke passed by at speed without a glance in our direction.' The Bevin Boys sitting patiently in the lecture room got little more attention, as Geoffrey Mockford remembers:

> At last the door opened and a shaft of sunlight flooded the room. The lecturer leapt into action with his pointer, describing an air lock below ground, which explained the slide that had been projected on the screen for a considerable time. The sunlight was interrupted by a large figure in military uniform and a fat face peered round the doorpost. It hung there for ten or 15 seconds and was gone. That was our royal visit.

The Duke did not venture down the shaft.

Throughout the period that Bevin Boys were trained, colliery managers barely coped with the responsibility. Under constant pressure for more and more coal, faced with labour shortages, union disputes and day-to-day pit problems, they were sometimes not even able to give the Bevin Boys the time they should have had underground, and a few groups were even sent on their way after only three weeks. Managers were at a loss as to how to deal with the recalcitrant among their trainees. The very worst were got rid of,[15] but men couldn't be sacked. Sometimes managers stopped a man a day's pay. But as trainees already had only shillings left after meeting their outgoings, putting them into debt only bred resentment. And sometimes, no doubt in exasperation, managers took such action in circumstances that trainees found unfair. Warwick Taylor and his companions one morning trudged through deep snow from the hostel built not far from the Welsh training colliery, arrived for an 8a.m. lecture ten minutes late, and were docked a day's money.

> It seemed pointless to remaining in the class for the rest of the day without being paid when we could be enjoying a day in the open or visiting the cinema in the town of Blackwood – which is what we did. That incident, plus the fact that we were taken off training and deployed shovelling snow off the colliery railway tracks, resulted in a mini rebellion that extended the training programme.

How valuable the programme was is an open question; indisputably the quality of it varied from centre to centre and at centres from time to time. Some Bevin Boys found the whole business a waste of effort. For Geoffrey Mockford it was 'ridiculous'; for David Reekie 'a farce'. For Gerald Carey it was 'badly organised and inefficient – and, it goes without saying, lacked discipline'. Owen Jones simply says: 'The so-called training furnished me with no abilities to work underground.'

Perhaps most men felt the training at least gave them an inkling of what miners called 'pit sense'. And some men go further. 'I still have my notebooks,' says Doug Ayres, an engineering student from Bloomsbury in London, 'and on re-reading them they are quite comprehensive, up to a good GCSE O level, and very detailed – we were told enough about shot firing and detonators to impress a modern terrorist.' Desmond Edwards remembers his lectures are being 'of a very high standard on all technical and human aspects – which contrasted with the sometimes dreary and ill-presented renderings at university.' And Ian McInnes, one of the relatively few men to stay on in the mining industry after his Bevin Boy days were done, comments: 'Based on what I saw offered to new entrants in both the Australian and American coal industries 40 years later, my training at Chislet in 1944 was much superior.'

*   *   *

Those seeing a colliery close to for the first time found it an intimidating place: the surrounding slag heaps[16] smouldering here and there like small volcanoes, the buildings black with coal dust, belched over by a giant chimney or two that made the air acrid. And dominating the landscape, the steel lattice headframes squatting over the shafts, their giant wheels whirring as coal was wound up from the pit in the cages.

All cages were dropped and raised on the steel hawser that ran from the winding wheels into the elevated engine room and ferried men and coal, though never at the same time. But they varied in size and construction: some were single-tier carrying two tubs at a time or a dozen men; others were up to five tiers deep. Two-deckers were usual, and when men used them those on the bottom generally had to crouch. Some cages were open-barred; others were of solid metal perforated with large holes. For stability, older cages ran on wooden battens attached to the shaft walls; most were steadied by lesser cables at each corner.

The thought of stepping into any cage to descend the shaft for the first time, in some pits a drop of almost a mile, was something that Bevin Boys constantly thought about with trepidation. The miners loved winding them up, just like the coal. 'You'll descend at a terrific rate,' the locals assured Roland Garratt, 'because "Mad Harry" is in control of the winding gear – and watching the rope [hawser] snatching and looping from the winding house it was obvious the cage was descending at the rate of gravity by special arrangement for Bevin Boys.'

David Reekie remembers the miners enjoying their discomfort and, a born parodist, goes on: 'It all depends on 'ow much ale Walter 'as 'ad last neet on 'ow fast you go down. But if tha' flex tha' knees and swallers 'ard, tha'll be there afore tha' breakfast.' He remembers some wag starting to say, 'Going down, ladies' underw-' and being cut short as the cage plummeted.

When cages carried tubs, full or empty, they were wound at 70ft a second; men were carried at 30ft a second – and enginemen prided themselves on being able to begin the descend and halt it so gradually that it wasn't possible to detect motion. Bevin Boys probably were given an 'initiation drop' at something like tub speed, and a couple of jolts by a few stamps on the brakes for good measure at the bottom. Not that the practice was condoned officially. 'The deputy had come down with us and he was furious,' says Dan Duhig. '"I'll deal with him when I get back up," he said. "No, bugger it, I'll deal with him now", and he got on the wind-up phone.'

Brian Evans' first drop, 'standing on an open-bottom cage with only the tub rails to stand on,' taught him 'to tuck the ends of my trousers into the tops of my socks to stop the wind from forcing my wedding tackle from finishing up around my neck.'

The cage's rattling and groaning, and what seemed to Geoff Baker 'to be an uncontrollable speed', brought an instinctive reaction from him:

> My hands sought that part of me that appeared most threatened. The cage slowed down and the lights of the pit bottom loomed into view. The sight of human faces! I was still of this world. At that moment I would have agreed to spend the rest of my life down there rather than go back up.

Harold Gibson had a worse first experience than most. In normal circumstances the cage slowed as it reached the bottom and the engineman applied the brakes – giving those in it the odd sensation that they were going up, not down. Gibson, however, was going to a tunnel halfway down the shaft and the engineman stopped the cage dead, leaving it 'bouncing up and down as though the steel hawser was elastic. Bungee jumpers must be out of their minds. When I got out my knees had turned to jelly.'

The change in air pressure, exacerbated when the cages passed each other like a piston in a cylinder, caused numbers of men, including the Cockney actor Harry Fowler, already well-known in films and on radio, 'terrible bleeding earache, like my eardrums had caved in. Swallowing helped like on an aircraft – it was far worse if you had a cold. One of the lads had a trickle of blood from the nose.' Other men left the cage and were violently sick. Owen Jones fainted. As well *The Times*' correspondent (who presumably never made the drop) wasn't around to discuss the thrill of it all.

Three

Learning the Ropes

B evin Boys expected to go directly down the pit once they got to the collieries. A few did. For the first two weeks – and sometimes more – when they thought they'd be underground finishing their training, the majority found themselves toiling on the surface. Colliery managers said the jobs they did were part of their training, that they were being taught from the bottom up (more accurately, from the top down). Bevin Boys looked again at their reporting instructions, which stated they were to be employed 'in or about a coal mine' – that 'about' now taking on a previously unnoticed significance – and concluded they were being used as cheap labour.

Whether that was true or not, they stacked timber, shovelled coal into railway wagons in the colliery sidings, and pushed tubs filled with waste to the slagheaps. At the Nunnery colliery in Handsworth, south Yorkshire, Alan Brailsford knocked down a brick wall and chipped mortar off the bricks. At the Homer and Sutherland near Stoke, Doug Ayres went into the firestone quarry adjacent to the pit, where he hacked away with a pick and shovel at the red mudstone, loaded it into a tub and pushed it to the brickworks. 'That quarry work was the hardest I have ever done in my life,' he says. 'It was winter and I had freezing water round my ankles. It was so cold my fingers were raw, and sometimes it rained and I carried on.' Many Bevin Boys got used to wearing an old coal sack on their head for protection.

Other Bevin Boys heaved tubs of coal out of the cages when they came up from the pit. That was relatively straightforward at the downcast shaft, which was open at the top for the intake of fresh air; it was less straightforward when the upcast shaft, from which the pit's foul air was dragged out by a huge fan and which was completely sealed, was also used for drawing coal. That generally wasn't the case, but it was at Littleton in Cannock, Staffordshire, where David Day had to contend not just with the tubs but with the airlock doors. When the cage rose, senior miners lifted the inner doors and pulled three tubs off each of the cage's two decks. Day's job was to stop them bumping through the outer doors before the inner ones were closed, which he did by sticking a 'locker' (a metal or wooden bar) in front of the wheels, 'or simply hanging on to the tubs like grim death'.[1] Then he pushed the tubs out – the sudden inrush of air, making the dust on the coal swirl up and blind him – as quickly as he could so that the inner doors could be opened for the next arrivals. At least Day escaped underground after a fortnight; Peter Allen, a trainee surveyor with the Great Western Railway from Bushey in Hertfordshire, spent three months at Ryhope colliery south of Sunderland pushing full tubs from the downcast shaft over the weighbridge for their weight to be recorded, then to the chute that delivered the coal to the screens for sorting, collected empty ones and brought them back to be returned to the pit bottom – work that 'had little to recommend it except I discovered muscles I didn't know I had'.

By common consent, 'the screens' was the worst job in coal mining, down the pit or up, normally the job of miners who'd been injured or were sick or were too old to go underground. It meant standing in an open-sided shed for eight hours separating out the stone or slate from the coal as it travelled along a moving metal belt and assisting it through different sized holes into the hoppers beneath.[2]

The noise as the coal from the chute hit the belt was deafening – worse at pits that used tipplers: cylindrical cages of iron mesh into which the tubs were manoeuvred, gripped by the axles and then revolved, sending the coal thumping down. At Gresford colliery near Wrexham in north Wales, John Wiffen communicated with his fellows in sign language because 'no conversation was possible, and for hours afterwards there would be a ringing in the ears and a sense of deafness in them which I thought at times might be permanent'. The billowing dust was almost as bad as the noise. Desmond Edwards, who worked on the screens at the Tower in Hirwaun, south Wales, remembers that when every tubful of coal descended, 'a dense dry cloud drifted along the belt and suspended light bulbs would temporarily vanish in the blackout'. Most collieries had spray or sprinkler systems to keep down the dust but they worked less often than they did. At Gresford the management handed out motorcycle goggles.

Sometimes the screens meant climbing into a hopper to help the coal through to the rail wagon below it. With coal clattering from above, it was a dangerous job. 'The hoppers had flat bottoms with a trap in them, not the sloping sides you see these days,' says Reg Taylor, who did the job at Gomersal colliery, 4 miles from his home town of Bradford. 'It really was a mad thing to have to do. And you couldn't breathe. I never got as filthy underground.' Perhaps it was appropriate that his colliery, its entrance in Nutters lane, was known as the Nutters.

For Phil Yates, a solicitor's clerk from Winchester who after his first four weeks stayed on at the Prince of Wales in Yorkshire, the screens 'was absolutely mindless – the most soul destroying job imaginable'. The disc jockey Jimmy Savile would agree. A miner from fifteen or sixteen, he was among the Bevin Boys at South Kirkby colliery in south Yorkshire. 'The screens,' he wrote in his autobiography,[3] 'is a job reserved for the young, the old, and the damned. It beats Hell because it's freezing cold.'

All Bevin Boys went underground eventually. Many, feeling they'd been exploited on the surface, expected indifferent treatment below – and were surprised, and gratified, to find that officials looked after them until they knew what they were doing; as many others elsewhere were unsurprised to be assigned to various stations and left to get on with it. 'I don't recall that either my friend George or I worked on the surface at Cossall [in the Nottinghamshire coalfield] or that we had underground training,' says Alan Gregory. 'We simply turned up on the first day and were allocated jobs.' Those who'd laboured on the surface were glad to get down into the warm. 'I spent five weeks on top,' says Doug Ayres. 'It was the thought of the warmth down there that kept me going.'

The routine of entering the pit, the routine of life belowground, was quickly established. Up the steps or the gantry to the bank, the built-up workings around the shaft. Tier by tier, the winding man raises the cage to the landing stage, the banksman thrusting in his chocks to hold it in position. Randomly he searches the men for cigarettes or matches – potential dangers in almost all pits. He removes the chocks. The cage drops. Later in the day, when coal is being wound, the cage will fly up and down the shaft every minute, the winding wheel a blur, like the propeller of an aircraft, the winding house wreathed in steam. One minute, two, three: the drop depends on the depth of the pit. There are always jokes, the same jokes. One from the Potteries, remembered by Ayres: 'Dust eer, they've paid won and threypence for this at Blackpool.' The cage slows. Men judge the winding man's skill, or his temper. They bend their knees: if the cage drops suddenly on to the chocks at the bottom they can damage a knee. If the winding man's braking hard, they're on their toes, ready if necessary to jump at the last moment. For some Bevin Boys the drop never gets any better. The rancid smell of pits clothes and tobacco on the jammed tiers don't help. Nor the rising smell from the pit: a smell like a doused coke fire, the smell of coal released from a million years of burial. These Bevin Boys leave the cage faint or dizzy. But some regular miners don't like the cage much either. Another joke, remembered by Michael Edmonds, from Bedwas colliery near Caerphilly: 'There's men going on the bond [cage][4] as won't go near the scenic railway on Barry Island.'

*   *   *

A pit bottom was like an unfinished underground railway station, the roof high, the walls whitewashed, the tunnels – always called roads or roadways – leading away through arched brickwork. The lighting in the most modern pits was electric, with air-tight installations, but most ran on compressed air, which gave off a constant hum. In most pits two sets of rails ran along the main road, one for empty tubs going towards the face ('inbye'), the other for full tubs returning ('outbye'). Some pits laid track with three rails that branched into four only at junctions, allowing 'ins' and 'outs' to pass. Yet other, older pits had only a single

track and when full tubs met empty tubs, the empties had to be turned on their
side off the rails to allow the fulls to pass, and then were turned back on.

In pits with mechanised haulage, the engine room housing the huge main
haulage engine, usually electrically driven, was at the pit bottom, often in
a whitewashed cave hewed in the rock, behind its own airlock. In pits still
hauling with ponies, the stables were nearby; by law, stables had to be in the
best-ventilated area.

In most collieries men had collected their lamp above ground in exchange for
their personal token, both stamped with their number – also their number on the
payroll. Under the system if a man went missing underground it would become
known when his token wasn't reclaimed at the lamphouse. Fewer other collieries
took the extra precaution of issuing each man with a second, different token,
also stamped with his number, this to be hung on a board belowground that also
showed his work station. Not only was the system quicker to show if he went
missing, it showed where he'd been working, too.

As men assembled at pit bottom – miners never used the definite article –
the overman[5] quickly handed out the jobs. Once detailed, men set off to where
they were supposed to be; and that meant walking, often for miles, through
progressively lower roads, with only their lamp to guide them – away from pit
bottom only the junctions (or landings) were lighted.

In a handful of antiquated naked flame pits, men were issued with candles to
guide them underground, as all men were in the century before. In the relatively
few 'naked flame' pits free of dangerous gas, men had carbide lamps;[6] cap lamps,
powered by a heavy battery attached to the belt, had started to come into mining
in the 1920s but their introduction was slow; the overwhelming majority of men
during the war still had the archetypal 'lighthouse', an electric accumulator lamp
weighing between 9 and 11lbs.

On the surface, they hooked this lamp to their belt, adopting a swinging gait
that involved a measured side-to-side roll to prevent the base of it from striking
the inside of their knees. According to Geoff Rosling, a Bristol schoolboy sent to
the Wyllie colliery near Caerphilly, 'Bevin Boys battered their knees black and
blue before they learnt the knack.' The gait, which had a certain machismo about
it, was irrelevant underground; whether 'lighthouses' were left on the belt or
carried, in roadways averaging 3 or 4ft, men moved in a stumbling crouch, in
single file, a small piece of card slotted at the back of the lamp's glass to prevent
the man behind from being dazzled. The coal dust under their feet 'was like black
velvet,' in John Wiffen's words. 'Inches and inches of the stuff, years and years
of it,' remembers David Roland from his time at South Normanton colliery in
Derbyshire, 'and you breathed it in from the people in front of you churning it
up.' Few Bevin Boys learnt to keep up with the regular miners underground, even
when they copied them and used a locker as a kind of walking stick. The miners
had an incentive to move quickly: they didn't start to earn until they were at their

workplace – and that could take an hour or even more to get there. It wasn't just the distance or the average height of the roadways: some pits had one-in-three gradients hundreds of yards long, which came as a shock to many Bevin Boys, including Brian Evans, posted to Beech Tree colliery at Lye near his home in Stourbridge: 'I didn't realise you have hills underground.' Some gradients were so steep and so low that the only way up or down them was scrabbling on hands and knees. 'And we didn't have kneepads,' points out Geoffrey Mockford, who was at Pleasley colliery, Derbyshire.

Men had to be mindful of their heads and their feet. 'We acquired the mining terms of warning,' says Mockford. 'A low area: "Head" as you approached it, which was passed down the line. An obstacle you could fall over, a roller for the haulage cable, say. "Feet" passed down the line.' Cap lamps had the obvious advantage of leaving the hands free, but, as Doug Ayres points out, 'Directly above the eyes they illuminated everything but with very few shadows and no sense of perspective – it was hard to judge distances. And if you stared ahead at things they weren't discernible, you had to move your head from side to side.'

Even the ventilation system presented obstacles. If the air drawn down the main shaft had direct access to the return shaft it would have travelled the short way out. To prevent that from happening, a series of pairs of doors, some of steel or solid wood but most of tar-covered sacking (brattice cloth) on wooden frames sealed roadways and directed the airflow through the workings. Vent doors had smaller openings in them for men to use. 'The secret,' according to Geoff Rosling,

> was to open the first of each pair and let everyone through, close it, then open the second. On no account should both sets be open at the same time. But frequently they were – and the air came roaring through with a whirlwind of dust, enough to blow off a pit helmet.

Some coalfaces were as much as 6 miles from pit bottom, too time-consuming to walk; in such cases some form of transport utilising the haulage system was provided part of the way: flat trolleys or adapted coal tubs without sides and fitted with timbered seats, but mostly ordinary tubs linked together in what was known as a paddy. Men travelled, at double the speed as when filled with coal, uncomfortably and even dangerously. The first time he travelled on a trolly, Roland Garratt, who was at Hanley Deep in Stoke, 'sat upright at the start as the roof looked around six feet high, but the miner I was to work with tapped me on the shoulder saying, "Lie down flat, youth, the roof gets very low." We raced down. The roof came to within inches of my nose.' Adds Geoffrey Mockford: '"A second-class ride is better than a first-class walk," was the miners' favourite comment.'

There was no chance of a ride back – tubs were full of coal and the system busy hauling it.

All pits were noisy places and noise reverberated from coalface to pit bottom and from pit bottom to coalface. Everywhere the roadways were a cacophony of sounds that overlay each other. Full tubs cannoning empty tubs out of the alternatively arriving cages with considerable force. The compressed-air engines powering the conveyors rising from a high-pitched whine as they started to a shriek. Coal crashing off the coalface chutes into the tubs. The never-ending rattle of wheels on rails. Bell signals jangling. Shouts. To John Squibb at the Shop pit near Stanley in County Durham, 'it was like some beast was bellowing from the depths of the earth'.

Other than at the inevitably draughty pit bottom and, at the diminishing end of the airflow, the coalface, which was invariably hot, most pits kept a constant temperature, winter and summer. The majority were dry, some were wet, some humid; others had a little of everything. Many of the pits in County Durham were shallow, 400–800ft, with the water table very close to the coal and so wet that someone was permanently engaged in ladling the water into a special tub that was then taken to a pumping-out point or the sump. There was often so much water coming off the roof in David Reekie's Yorkshire drift that the candles which men had instead of lamps couldn't be kept alight 'and the tubs sometimes floated, even the full tubs floated. Moving an empty tub was more suited to the skills of a bargee.' In the same county, Alan Brailsford's drift often had water pouring in. 'We wore waterproofs but got soaked anyway and became very tired ploughing through the deep mud in our huge safety boots. The smell was disgusting, too, sulphurated hydrogen, bad egg smell – I can still recall it.'[7]

Later, Brailsford was transferred to a deep seam 'down a steep incline which became hotter and hotter as you went down. We shed our clothing as we went.' Peter Allen, a Great Western Railway trainee surveyor, also removed his clothing as he went towards the coalface at Ryhope colliery near Sunderland. 'We had caches along the route. By the time we got there we had a jockstrap or not even that.' Brian Folkes, a trainee chartered surveyor from Buckhurst Hill, Essex, transferred from Durham (South Hetton, 800ft) to Kent (Betteshanger, 1,900ft) and found it not just wet but humid. 'Shirts had to be removed on leaving the shaft to walk inbye,' he says. 'Even then rivers of sweat quickly ran down face, back and chest. But the worst problem was water penetration; we were under the Channel and sometimes we thought we were in it.'

The temperature of strata increases one degree Fahrenheit for every 92ft in depth – and large quantities of water were taken up in the warm circulating air, resulting in humidity of up to 90 per cent. Comments Ian McInnes, who spent his service at Snowdown colliery in Kent, at 3,030ft the deepest in the country:

At the coalface and in the main return airways, your sweat couldn't evaporate off your skin and merely ran down your body. If you had clothes on they were soon soaked. The only solution was to work starkers, wearing your hard hat, your lamp belt and your boots – no socks, they would collect all the sweat that ran down your body.

'The temperature was always about 95°F and on one occasion when the ventilation was poor it was more,' according to Jim Ribbans, who transferred from the Welsh Werfa Dare colliery near Aberdare to Snowdown. 'Men were being carried out.' Once on night shift at Hanley Deep in Stoke when a set of air doors jammed open for an hour and a half and the airflow went straight up the return shaft. Like the rest of the shift Roland Garratt continued to work with hammering chest, and 'the dust just stayed suspended in large globules through the roadway, just as if each globule was hanging from the roof on different lengths of invisible wire'. In humid pits it wasn't uncommon to lose half a stone on a shift.

Whatever their training had led them to expect, most Bevin Boys hadn't anticipated what a working pit would truly be like. Looking back, Geoff Rosling realised that the real thing bore no relation to training's 'highly sanitised demonstration world'. 'They obviously didn't want to scare us and gave us the impression that pit life wasn't going to be so bad after all,' says Ronald Griffin, who'd worked in a small electrical repairer's in Smethwick near Birmingham before call-up and who went to the Aberbaiden pit near Kingfig Hill in south Wales. It amuses him to remember one instructor, 'a funny little white-haired man who kept quoting Charles Dickens, who by the eloquence of his voice led us to believe we were about to enter an Aladdin's cave.'

*   *   *

Bevin Boys who'd anticipated the pit being uniformly warm were to be disappointed if they were employed at pit bottom. Here the inrushing airflow was stiff and here, in pits that had mechanical haulage, gangs of men uncoupled sets of full tubs from the cable – always referred to as a rope – as they came outbye, or assembled sets of empty tubs and sent them inbye, or pushed full tubs down an incline into the tiers of the cage, or on the opposite side pulled out empty tubs – and whether pushing or pulling, twisted the tubs on the flat steel plates that formed the pit bottom floor.

Most Bevin Boys started pit life working at pit bottom; some spent their entire service there though most also did a number of other jobs. Geoffrey Mockford has never forgotten his pit bottom days: 'It was the coldest place I had ever been. There were some working there who weren't Bevin Boys and they were wiser than us conscripts – they had coats, scarves, even gloves. We in our ignorance had none of these things at first.'

It was unremitting work on the 'flats', studded pit boots scrabbling on the steel surface, always slippery with condensation, or rain or snow on the tubs from the top, or with grease from tub wheels. The pit boy waiting for the 'empties' to descend was often soaked when water in the shaft sump, which was 20ft deep, was overflowing the massive timbers and a cage splashed down rather than landed. It was unremitting work whatever Bevin Boys did at pit bottom: as many as 400 empty and 400 full tubs came in and out of the cages every hour. Ken Sadler, remembers being on the gang that uncoupled sets coming outbye at Blagdon colliery near Newcastle.

It was non-stop graft and it was hardest when I became 'tailend boy'. My job was to leap on to the bumpers of the last tub, crouching to avoid overhead girders, and disconnect a cotter pin holding the tail rope on the tub. That released the rope and allowed the set to run free towards the shaft, where others would stop it by other sticking dregs [lockers] in the tub wheels. I'd run down to help with uncoupling the tubs and delivering them to the onsetter,[8] who'd push them into the empty cage. Then I'd run back. Throughout the shift full sets kept arriving from different districts. There was no letup.

Gangs of Bevin Boys were deployed further into the pit doing a similar job. Because underground distances could be great, or because roadways changed direction, small, stationary engines were required to continue the haulage-rope system. These were at junctions, where tubs were transferred from one moving rope to another, twisted on and off over yet more steel flats. The work here was no less unremitting – long sets of empties had to be broken up into smaller units and sent to the coalface, possibly to different districts and even different collieries that were interlinked. But there were compensations if they weren't close to the coalface. Besides having lighting, junctions, being on the main roadways, tended to be airy rather than draughty. And most were at least of stand-up height.

Not all jobs on haulage were frantic; some, indeed, were so intermittent that time hung heavy. At Welbeck colliery in Mansfield, Owen Jones was deployed in a desolate part of the pit and his job he found as soul-destroying as the screens.

I stood alone at the back of a long line of stationary tubs and rarely saw anyone. I waited for the empty tubs to be sent up from the pit bottom. When they reached me, I unhitched them, coupled them to the last stationary one and hurried to the end of the run of 24 tubs to go through the whole thing again. Nowhere to sit. Just stand there waiting for the next run. In sheer boredom I chalked sentences on the tubs from the classics and other books. I always loved books and all my life have tended to see people as characters from what I read. This became a matter of interest. I heard the characters being discussed in the showers and it amused me because I'd told no one I was responsible for this graffiti. Then one day when I surrendered my lamp I was told to see the manager. He told me to stop as it was confusing the checkweigh man, who had no interest in 'Barkiss being willin' [*David Copperfield*] or Elizabeth Bennett's relationship with Mr Darcy

[*Pride and Prejudice*], but only the source of the coal coming up in the cage to ensure the appropriate coalface workers were paid.[9]

There were tasks that were possibly more mind-numbing. A Bevin Boy could find himself posted along a roadway where tubs had a habit of jumping the rails or where the endlessly moving rope occasionally jammed in the pulleys. It was a lonely existence, in the dark, hour after hour, with only a lamp for company. Mostly, nothing happened. When it suddenly did, derailment or jam, the job then was to get on the bell signal fast and get the engine stopped – and sort out the problem. Almost anyone working in a pit could be confronted with a derailed tub, or a dozen. Empties were one thing, fulls quite another. Besides having to shovel up scattered coal, men, sometimes on their own, had to lift the tubs back on to the rails, first one end, then the other. Some men couldn't do it. John Marshall, who until call-up had been an RCA sound engineer working and living in the West End of London, 'wasn't physically up to the job' at the Stargate colliery, Ryton, near Gateshead. 'The first day I saw a tub come off I said to one bloke, "Who's going to put that bugger back?" and he said '"You". I just wasn't able.' 'I wasn't strong enough,' admits Ken Sadler, 'and I didn't have the miners' knack of bracing the legs against the side or lying on your back and using you legs.' Peter Allen was another: 'The experienced miners, small as most were, could lift a fully loaded tub back on the rails. I never ever mastered the technique. I usually needed a crowbar and the help of another Bevin Boy.' Old-time miners had John Wiffen's admiration for the way they could put their back to a tub and made it jump on to the rail 'like a thumb with a tiddlywink'.

Norman Bickell wasn't lonely in his first job at Monckton Pit Gates near Royston in Yorkshire: he was working in a busy area close to the coalface. But he was bored. His 'exalted task' was to chalk a number on each filled tub as it left the coalface.[10]

After two weeks I'd had enough. You can't imagine anything more mindless. We were producing 380–400 tubs each shift. So I'd start off at one but then get to about 50 and go back to 48 and miss a few every so often. It wasn't long before I was relieved from this post. Never was any good at figures – well, that was my excuse.

The best job on haulage was 'driving' one of the small stationary engines and Jim Bates did that for six months at Hem Heath colliery in Stoke, 'hitching and unhitching wagons [tubs] to a main and tail rope[11] and, with a ringing of bells, sending them on their way.

I was in a little world of my own, untroubled but by runs of wagons sent to me to deal with every so often. I found that with a cap lamp I could read, and I read Tolstoy and Hardy, two of my favourites, and since I was studying to become a lay preacher, theology and the Bible.

I had a pressure gauge on my engine and people were constantly ringing up to know the pressure if they weren't getting what they needed, asking what it was reading. One day the under-manager passed by and asked what the reading was. I though he said 'What are you reading, Jimmy?' and I replied, 'The New Testament.' But I'd misheard and his reaction was understandable. All around the pit it went that Jimmy, instead of getting on with his work, was reading the 'bloody Bible!'

Arthur Gilbert got an engine driver's job at Berry Hill colliery, also in Stoke:

It was the practice to stand to manipulate the various levers, but after a few days I manufactured a bench seat, which as days passed acquired a backrest and finally armrests. With a cushioned seat of layered brattice, I was able to manipulate the engine more or less lying down.

Doug Ayres went one better. Operating an engine on nights – when coal wasn't being drawn and a skeleton crew was cutting a new roadway – he received a signal only every two or three hours to pull out a set of tubs full of rock, so, that dealt with, 'I would lean back, close my eyes and have a nice sleep. A jolly pleasant few months – during which I could get about during the day.' Perhaps it was recompense for an earlier job – as the 'twister' on the steel flats directly adjacent to the loader end, one of the most strenuous jobs on haulage and only marginally less dirty. 'You could write your name on my chest,' he says. As to the twisting, he adds: 'You got to know exactly how a tub would move and turn, so you knew the right moment to slide it over with the minimum effort. But effort enough. I speak with feeling. I often did 600 tubs a day.'

Arthur Gilbert lost his cushy number because he got cocky.

I became so proficient I started to use my boots on the levers. One day the inevitable happened. My foot slipped and I wasn't able to stop the loads coming down the steep incline in time. The tubs were dragged over the rail and collided with the engine. The resultant tangle of overturned tubs and their coal caused a bottleneck and stopped the whole output of the district for a considerable time.

Next shift he found himself at the loader end of one of the 'gates', small tunnels at right angles to the face, often a mile or more long, down which the coal came on belt conveyors. Whoever did this job crouched with empty tub after empty tub underneath the lip of the belt, which from a height of 6ft streamed down the coal. When each tub was full, he manhandled it out and dragged in another.

'The continuous fall of coal resulted in a constant cloud of dust and clogged my eyes and nose,' says Gilbert. 'The tubs filled at an amazing rate and I didn't stop pulling empties, turning them, pushing out loads. The harder I worked the deeper I breathed and the more dust I inhaled. At the end of a shift I was black as the ace of spades.' Roland Garratt's description of what being at the loader end was like is graphic:

> The coal came off the conveyor belt more like high pressure water pouring out of a broken main. At times the excess coal fell on the rail lines behind the loading head, making it impossible to push tubs underneath the belt – then the coal just poured off, filling the roadway. We were climbing over the pile in an effort to switch the belt off, which brought a torrent of bad language from the face.

There were always arguments among Bevin Boys as to what was the dirtiest job in the pit. The loader end certainly qualified as one, especially at the start of a shift: the mechanical coal-cutter[12] that the night before prepared the face for the day shift produced vast quantities of dust and for half an hour nothing but dust discharged from the chute. A job that was as dirty as the loader end was being in what was called the spill-hole beneath it, and if anything was a worse job. 'Large amounts of coal would go over the side of a tub,' says Geoffrey Mockford.

> It was the poor unfortunate in the spill-hole who had to keep it clear, shovelling for all he was worth. I had my fair share of it and it was the hardest work I did in the mines. The size of some of the lumps made shovelling impossible and there just wasn't room to heave them out, even if they weren't too heavy to lift. Breaking up lumps took time and all the while the hole filling. When the loader clogged you weren't popular with the colliers. Stoppages cost them money.

The shallow troughs or belts running parallel to the face on to which the colliers shovelled their coal and which fed the loader could also jam and Bevin Boys trying to keep clear the drums at either end, often shovelling on their knees, could expect a volley of oaths when it did. Geoff Rosling:

> The problem was the belts were usually in poor condition with quite large pieces missing – materials were scarce during the war – and quantities of smaller coal fell through and carried down the track to the drum. I shovelled like fury but occasionally the drum did seize and stop production, and I was showered with far from affectionate comments. The engineers would show up, reverse the belt to clear the drum and work would restart. I was sometimes very close to tears of frustration and pain.

In intense, constricted conditions, men swore. The majority of Bevin Boys got used to being sworn at round the coalface. It rolled off the back of Reg Taylor when he was helping the shot firer hand-drill the face; and of Donald Whittle, who after training at the Newtown pit near Manchester stayed on there to work, where one of his jobs was keeping the colliers supplied with wood or steel props throughout the shift.

Strictly speaking, Whittle, from nearby Eccles, who'd arrived in the pit after a year reading history at Oxford, was 'the powder monkey, the fireman's [shot firer's] lad who carried the tins of gunpowder'. But the other part of his job took most of his effort,

> pulling and heaving in a seam three feet six and often much less, hoping I heard what the colliers were yelling. Sometimes I got it wrong – the blasting often left me with ringing ears; there was the engine noise, too, and the heavy dialect. When I did it added more words to my vocabulary.

What stays with Taylor isn't the colliers' vocabulary but the hard graft.

> The drill had a handle and lever, each with a ratchet on the end of the drill shaft. This was held by an adjustable stand in a position parallel to the roof. There wasn't enough room for the handle or the drill to make a complete revolution. So we had to push and pull alternately until the bit had penetrated the stone to a sufficient depth. It took a hell of a lot of effort. And after the bang I shovelled the pile of shattered bits into a tub and pushed it to pit bottom.

In his Staffordshire colliery, David Day helped the man who drilled the holes with a big, powerful compressed drill called a tadger, which might appear to have been a better option than Taylor's but had its drawbacks. The tadger was connected by a long rubber cable to the power in the middle of the road; the job of the driller's mate was to see there was enough cable and that it didn't get snagged. But as the driller moved along the face to where a colliery wanted his stint[13] drilled, the cable caught under the fallen coal in other stints and, scrabbling to free it, Day set the coal tumbling out of one man's area into another's, making him the recipient of more colourful language. To him the job was 'an ordeal, a humiliation, a nightmare'.[14]

Heat, noise, the swirling dust that enveloped the entire coalface area: Doug Ayres watched his fellows, often no more than silhouettes, their lamps dimly glowing but emitting no light, and thought 'this the nearest thing to Blake's vision of Hell'.

*   *   *

Pits without rope haulage or with rope haulage only part of the way from pit bottom to the face, relied on older methods: hand hauling or ponies (and in the very few pits with seams high enough, cob horses). In pit parlance the job was 'putting' (pronounced as in golf) or, in Yorkshire, 'tramming'.

Hand putting was exhausting work: putters brought tubs from the head of the mine to the coalface, often helped fill them by shovel, then pushed them back along the rails, returning with another empty. Geoff Darby, a trainee engineer with Huddersfield Corporation waterworks department, did the job at Lepton Edge colliery near his home town in a seam that was only 31in, pushing with his head between his knees. 'It was punishing,' he says. 'It was almost impossible to keep up – so more bad language.' Non-mechanised pits tended to be less well maintained than mechanised ones and tubs would frequently catch on bent propping timbers or lodge under the roof. Peter Allen was often frustrated and miserable in such circumstances: 'A tub off the rails, wedged, no room to get round or over, and hewers [colliers] waiting on you. Imagine . . .'

'Bringing a tub containing a quarter of a ton of coal or waste away from a collier's "gate" along a roadway three feet high was no joke,' comments David Reekie, who was at Horse Riggs colliery at Morley near Leeds, where the seams averaged 18in, down to 14 in places, and colliers lay on their side, using a short-handled pick and shovel, and threw the coal behind them. At times Reekie crawled in to help, scooping the coal out behind him in turn and then crawling out backwards to fill his tub.

Then just to add spice to the job, there were occasions when there was a shortage of incoming empty tubs or the trammers had too many colliers to look after. On arrival back at the face there'd be two or more full tubs waiting. 'Tha' can manage two if tha' tries, lad', they'd say.

Older tubs, their sides worn down by several inches, were easier for the putters to handle and, though they held less coal on the weighbridge, were quicker and easier for the colliers to fill. And, Reekie discovered, there was intense competition among the putters for them.

If you were first in the queue, the first tub on the train [set] was yours, no picking and choosing. But by guile and stealth, a small tub could be lifted out of sequence from the train, provided no one was around, and whisked away and hidden in a disused gate.

Bill Hitch, a market gardener from Haddenham near Ely in Cambridgeshire, was on haulage at Kibblesworth colliery in County Durham when the manager asked him to go pony putting.

Men were complaining the putters weren't getting the work out. The manager came round and he says 'Bill,' he says, he knew everyone's names, 'take that pony and come with me and put a tub on and see if you can get in and out to the face.' I did and he says: 'Tomorrow morning call at the office, you're on datal putting' ['day' and 'telling' or reckoning – employed on a day-rate of pay]. He was so pleased with my work he gave me a 15 shilling bonus, which was a hell of a lot of money.

I had a good putter mate, another Bevin Boy called Bobbie Cave, who came from Liverpool – two putters to take out six men's coal. The first day there was talk of those six men would never get their work out with two Bevin Boys putting off them. Bobbie and me made our minds up that we would disprove this. Each hewer filled 12 tubs, so Bobbie and me had to putt 36 tubs each. And we did, plus three full tubs that went to our name and were worth a couple of bob apiece. We won the respect of all the men in the pit, but it was hard. People can have no idea.

There were nearly 23,000 ponies working in the pits during the war years as well as around 5,000 cob horses.[15] They were well looked after by full-time stablemen, groomed, well fed and, by law, had fresh water piped from the surface; some of the bigger pits even had proper baths. They came up into the daylight only during the pits' one-week summer break and, as George Poston saw at Handsworth Nunnery near Sheffield, 'it was joyful to watch. They came out of the cage nervous and trembling, but once in a green field kicked up their heels and galloped with delight.' Stablemen were known to give up their holiday to keep an eye on them.

Animals working in drift mines were luckier: at the end of the working day they walked out with the miners to their stables on the surface.

Dennis Fisher became a pony putter at Chilton colliery near Bishop Auckland but first he worked in the big underground stables, one of the few Bevin Boys to have such an experience. 'I loved the ponies, they were great,' he says.

We had all kinds. We used to get a lot shipped over from America and in one shipment they shoved in a mule to make up the numbers. He was a funny looking thing with a large head plus long ears which stuck out forwards. Peter and Paul were my favourites, and Titch, a little black one, that small he passed between my legs – his working days were spent in the stables putting the water tub to all the stalls and the empty tubs for the manure.

Peter and Paul had adjoining stalls, the best of marras [mates]. Peter was from Russia, ginger in colour with a dark brown stripe down his back to the end of his tail, a big, hefty, sturdy pony. Paul came from Scotland, running wild. A German bomber had dropped his bombs and a lump of shrapnel left Paul with a nasty scar on his hindquarters. He didn't like that being touched – he'd try to bite you. But he was a good worker. They were both good workers.

I cleaned out the stall ever day. No one would sit next to me on the bus home, with good reason. But the stables were a nice warm place to be as most places at Chilton were either cold or wet or both. I saw a miner punch a hole in a tin of soup and push it halfway into a tub of hoss muck. By bait time [meal time] the soup was red hot and he had to use his cap to hold the tin.

More than one stableman grew mushrooms in the underground stables on piles of manure.

A Bevin Boy who pony putted fetched his charge from the stable at the start of a shift and harnessed him; protection was afforded by a leather head guard to prevent the animal stunning himself on low roof supports or hanging chain pulleys, and a strong metal link like a giant paperclip suspended from his harness enclosed him from front to rear in case he was caught between moving tubs. Once a pony was ready, his putter walked him along the return airways – the main roadways were too dangerous – to where they'd be working.

Ponies had their idiosyncrasies. Syd Walker remembers three he putted at Great Wyrley pit in south Staffordshire:

Dot wouldn't start work until he'd had a roll in the dust outside the stables. If someone harnessed him up before he had his roll, he wouldn't budge. Those in the know wouldn't say anything if a Bevin Boy was given Dot to harness up before his roll. Jerry was a pony with all bad habits. He bit, he kicked, he had bad breath and the other end was worse. When you had him hooked on to a tub, you had to sit on the limbers [shafts] with one foot on the rail track and the other on the tub. An awkward position, coupled with the fact that the roof was low and you had to keep your head down – near the pony's rear. Jerry loved to stop in a really low place and pee and fart together, with the air in the pit coming towards you. Charlie was my favourite, dappled, with a long tail. His only fault was that he had only two speeds – stop and full speed ahead. At times he scared me out of my wits.

'As soon as a pony realised you were sat on the limbers as often as not he'd tear hell for leather down the way,' says John Squibb.

If this happened to be downhill and a low roof into the bargain, it put the fear of God into me. Sometimes the ponies were frustrating. They'd defy all your attempts to influence them to go in a certain direction. Or they'd seemingly deliberately pull the tubs off the rails at a curve.

Or, as Jack Garland knows only too well from days at Markham colliery near Caerphilly, stop in a the most inaccessible sections of the roads, out of sheer bloody mindedness, where the roads were hardly an inch wider than the tubs, and you were behind the train [set] and they were in front. Shouting at them was of no avail and you could only get them to move by heaving stones.

Like most Bevin Boys, George Poston tried to fool his pony as to how many tubs it was pulling:

On many occasion I tightened the chain links between the tubs before moving off. He overcame this by backing, pushing the tubs together with his rump before going forward. If the noise of the tightening links was more than two on a full train or more than four on an empty one, he would stop dead and nothing could be done to move him. In the course of time I discovered this pony had been down the mine one and a half years longer than I'd been alive and he knew his union rules.

Ponies knew the signals for the end of a shift, too, and if the putter wasn't careful would find he'd been left on his own, as Norman Bickell, who became a putter after showing his ineptitude on tub-chalking, was the first time he went out with Dandy – and 'he took off with my lamp, leaving me to walk back along the ventilation road in darkness. But I was very relieved to see him in his allotted stall – the one with the name Dandy printed on it.'

Ponies were clever in many ways. They were able to unscrew a man's water bottle and swig the lot. Or steal a man's sandwiches, which were carried in a closed tin box to prevent them going stale, by dropping it on a rail to make it spring open. 'If I left my food tin in my jacket hanging on a pit prop Jerry would tear the pocket and stamp on the tin to get at what was inside,' says Syd Walker. 'I always had to hide it somewhere out of his reach. Charlie didn't do anything like that. He was always grateful if I shared my sandwiches with him.' Most Bevin Boys did share, whether they putted the ponies or not. So did many miners, who brought down treats like cabbage stumps and fallen apples. Most men held the ponies in great affection because they brought laughter as well as exasperation. Some refused to lift a hoof until a titbit materialised. Some chewed tobacco. Others had a party trick of jumping up on a moving conveyor and running on the spot – and at the end of the shift one was known to hitch a conveyor ride towards pit bottom. 'I enjoyed working with the ponies and would willingly have done it on a permanent basis instead of just occasionally,' says Derek Thompson, who went down the Bedlington pit north of Newcastle. 'If you like animals you couldn't help liking them; most were nice, gentle creatures.' Geoff Darby took a shine to the ponies in training in Yorkshire and looked after them when he could; he was disappointed there weren't any at Lepton Edge where he went to work. 'The seams weren't high enough. They had Bevin Boys instead – they were cheaper.'

Dennis Fisher makes much the same joke: 'If there was a closing[16] the first thing management would ask, "Is the pony safe?", never mind the Bevin Boy. They could always get another of them off the government for nowt.'

*   *   *

It was said that the army needed twenty men behind the lines to keep one at the front. Coalmining needed a similar ratio of workers behind each collier on the face; colliers were mining's frontline troops.

Few Bevin Boys wanted to be on the front line or, indeed, had the capability, the mental and physical endurance, to work in the most arduous and dangerous place in the pit. A number, however, did. One was Dennis Fisher. He might have been disenchanted at having been called down, but mining was in his blood and, he reasoned, 'if you're down there, the face is the place to be, it's what mining was all about'. He had a low opinion 'of a lot of Bevin Boys who had no intention of moving any further than the shaft bottom', but has fond memories of a Bevin Boy called Claude who worked on haulage when he himself moved to pony putting.

> If you can find any wartime pitmen from Chilton, and there's still a few of us left, of all the Bevin Boys Claude is the best remembered. He had a jet black mop of wavy hair, he always carried a mirror and comb in his top pocket, and he was the only pitman that I ever knew who always carried a white handkerchief. When I was pony driving in the district he worked, we were under the marketplace in Ferryhill and when shots were fired they blew the candles out in the Catholic church. The complaints had the district closed, never mind how much coal was needed for the war effort. I never saw Claude after that.

Fisher exulted in his ability to fulfil his daily quota on the face, wielding 'a pan shovel whose size would make any ordinary man weep', and took pride in his pit reaching its weekly one.

> Each Friday, the first thing we did coming out of the shaft, we younger ones, was to look up at the pithead pulley wheels to see if the Union Jack was flying as then we knew the pit had reached its target. It always was in them days – they never bothered to take it down until it was flying in tatters. But we looked to see it was there anyway, and stuck out our chests.

Alan Gregory did various jobs in his pit (Cossall, near Ikleston, on the Notts/Derbyshire border) but 'always aspired to be a stripper [collier]' because he relished the challenge. He'd been looking after a set of shaker pans, stopping and starting them to order and, from time to time, 'when everything was going well, sliding over to lend the strippers a hand – it was a very good job from which to see how they went about their work'. Finally he got his chance as a 'day man', taking the place of any regular collier who failed to turn up. For the last nine months of his service, working a 10yd stint, he moved 15 tons of coal a shift, on his knees all the time. He makes light of it: 'It wasn't as tiring as it sounds – the edge of the shaker pan was only about a foot above the floor.'

Bill Gibbs went on the face at Clipstone near Mansfield because 'I decided a little more cash in my wage packet wouldn't come amiss'; he was the only Bevin Boy there to do so. David Roland at South Normanton in Derbyshire was similarly motivated. 'The wages were absolutely marvellous, the same as anyone's working down there for twenty years,' he remembers.

The barrister John Platts-Mills was another Bevin Boy collier, but he went on to the face through altruism. He sought agreement from the manager, heard nothing, but was so well liked by the men underground that he started anyway. When this was discovered he was ordered off but had support from the men and stayed.

> While working on the face, I hoisted eagerly my stint of ten tons a day onto the belt. This greatly amused the regular chaps, who found my eagerness for Ernie Bevin's need for coal quite misplaced. I would finish my stint, then help the fellow next to me. The older miners . . . stood round to watch the scene with amusement . . . 'Woud'st tha' like little brush and pan to sweep up dust and put on t'belt? Leave a bit for us when tha's gone back to tha' cushy job in London.'[17]

Platts-Mills never discovered what the objection to his working on the face was in the first place. Bevin, surely, had nothing to do with it.

The Durham and Northumberland coalfields were the least mechanised in England, though some collieries there had pneumatic picks that ran on compressed air – 'windy picks' as Dennis Fisher describes them. But Wales was the least mechanised coalfield in Britain, and while a fair proportion of pits had some rope haulage, the coal in almost all was hand got. In England and Scotland Bevin Boys had to be over twenty-one and working underground for six months before they could be considered to become colliers. In Wales, the part of the country with the greatest manpower shortage, a fair number of Bevin Boys, whatever their age, went straight on to the face.

Those who did worked with a 'butty', a partner who was a senior collier, and together they did all the jobs that different men did in mechanised pits: they hewed the coal, shovelled it into the tubs (called drams in Wales), laid their own rails, cut their own props, packed the space behind them with dry-stone walls to support the roof, and hand-drilled the coal or rock for shot firing – 'sometimes powerfully supported in the back till one's sternum felt like cracking', remembers

Desmond Edwards. Whether the work was less hard than in a mechanised pit is debatable; it was simply different, carried out to a different rhythm. As his instructor at Oakdale told Michael Edmonds, who fetched up at the Tower colliery near Hirwaun where Edwards also was, 'Machine mining means a long straight coalface but less art.'

Deryck Selby (the Empire, Cwmgwrach), a mechanical engineering apprentice from Walthamstow certainly found his first day arduous:

> We began the task of shovelling a ton of coal into the tub. Being right handed I placed myself strategically to throw the coal over my left shoulder, but my butty had different ideas as that was his chosen position. When the tub was filled, he scribbled some hieroglyphics on its side with chalk. Then Dobbin, 14 hands of solid muscle who'd brought the tub, reappeared. I can't remember how many tubs we filled. At the whistle the horses galloped off to their underground stables unattended. The men climbed into the empty coal tubs for the return journey to pit bottom. I was completely knackered. I couldn't match Jack North's strength. It was months before I could keep up.

Like every man who hewed, his physique improved. As did David Roland's: 'I became tremendously fit. I had great upper body strength.' George Poston 'became so muscle-bound I was unable to scratch my back'.

Frank Pratt arrived on the face at the Tower still smarting that he'd been refused compassionate exemption to look after his mother and made it clear on his first day that

> I was reluctant to work. But my butty, one Trefor Jones, soon persuaded me that before he could earn any bonus he had to earn my wages in addition to his own and as he had a large family to support he intended I should play my full part. Trefor was a man of such physical build that I was soon in the right frame of mind to contribute to the well-being of his family.

Other Bevin Boys elsewhere in Wales also got into first day confrontations – over purchasing tools.

Whether Bevin Boys were supposed to buy their own tools was something that was never resolved in any part of the country. The first Bevin Boys who'd staged a walk-out at the training centre over wages had been told not only that they had to buy their own picks and shovels but also their lamps and helmets. They'd refused – and officials backed off, diffusing that part of the protest. But there was never a government line on the issue and some collieries from time to time told Bevin Boys they were being charged for items, and while again some refused point blank, others paid, however reluctantly. The practice of trying to charge was widest in Wales.

Dan White, sent to Western colliery near Nantymoel and already taken aback that, 'over six feet and big with it', he was to be working in a 2ft 6in seam, was furious when told he had to buy his own tools. '"Like hell I will," I said. "A soldier doesn't buy his own rifle."' Similarly told at Aberbaiden drift mine that he had to buy a shovel, Ronald Griffin also said no.

For a start I didn't have the money, six shillings, I think. I went down from the surface on the paddy, a sort of iron seat on wheels which on a cold morning was enough to freeze the blood, packed with men sitting back to back for a mile or two, and then transferred into coal tubs for another three or four miles under the sea to the coalface There the doggie [foreman] took me to work with a collier named Selwyn. When he'd done that he asked me where my shovel was. 'What shovel?' I said, thinking I ought to have collected one on the way in. 'To work with. You can't work without a shovel, can you boy?' Honestly, I wasn't sure who was the stupid one, me or him. Selwyn lent me his spare. I never did buy a shovel but continued to use Selwyn's. To make up for it I slipped him the odd packet of twist.

Ernie Jefferies tells much the same tale from his experience at Clyncorrwg, though his reaction was much more aggressive:

First day taken underground by fireman and introduced to Dai Edwards, 65, tough as old boots. They spoke to each other in Welsh and the fireman departed. Dai, man of few words, gave me one of his two shovels and as he brought down the coal with his pick, I start shovelling. After about two hours the fireman returns – speaks to Dai in Welsh, then says to me 'Do you have a shovel?' 'Yes,' I say. He then repeated the question. I replied with expletives. 'What you think I've got here?' 'That's not your shovel,' he says, 'you have to buy your own tools or you can't work in the mine.' More swear words to the effect that that suited me fine. At first I thought he was taking the piss but he bloody well meant it. To say I was angry is an understatement.

Okay, I had a quick temper. But they didn't understand the situation. Young Welsh guys went into the pit as normal practice and like a carpenter on a building site they had to have their own tools. But we were conscripts, you know what I mean? So, interview with manager. 'I must contact the Ministry for clarification', he says. I never did buy a shovel, or a mandrel [a pick with one pointed and one chisel end]. Dai cut his hand too badly to continue work so I was attached to another miner. My reputation had preceded me. Yeah, I was a quick-tempered Londoner, but a hard worker; his tools were mine to use.

*   *   *

There were numerous jobs underground not directly involved in the remorseless circle of haulage. Most were general labouring – 'snigging' – bringing bricks, girders, pit props and other timber down the upcast shaft (coal was being wound in the downcast) and bringing them to where needed in the pit. All these jobs took Bevin Boys through the return air roadways where the ventilation was at its worst, the roof supports and track in poor condition, and water lay in pools. Some of the returns were so low that the only way to negotiate them was shuffling on hands and knees and, says Syd Walker, 'they were so thick with blankets of cobwebs you had to cut your way through them with a shovel'. Steel, scarce during the war, had to be salvaged for recycling from old workings, bent ones, where possible, straightened in an hydraulic press; Brian Folkes at South Hatton once recovered some pipes, which were too long for the pressure chamber between doors,

> so we had to open the slot in both at the same time and pass them through. This created a man-made hurricane. Not only were we covered in dust but the unlucky fellow receiving the pipes (guess who) had years of rust scale scoured from their insides blasted at him.

Another job that Bevin Boys who did it remember ruefully was stone-dusting. In mechanised pits the coal dust that came off the face and was blown along the roadways in enormous volumes was in very real danger of combusting if it wasn't coated with copious amounts of limestone crushed to the consistency of flour.[18] Jim Bates' first job in Staffordshire was 'as a "dustman", not collecting refuse but scattering dust.'

> Large paper sacks of limestone dust, about the size of a cement bag, came down and we had to scatter it everywhere, every ledge, every crevice, the tops of every girder in every part of the pit. The really hard work was getting it into the back air roads. You sweated and toiled, wriggling through narrow passages dragging a bag, which was very heavy and which inevitably tore, losing some of its contents. Then having got it where you wanted, you scattered what was left on roof and sides, whitening and lightening the road as well as making it safer. The other miners cursed us because the dust got into their machinery and clogged the rails. But we were a necessary evil.

'You threw the stuff up in the air to cover the high places and smothered yourself,' says Roy Doorbar, who was sent to work in the pit in his own Staffordshire village. 'You tried to get on the air side of what you were doing, but stone flour was up your nose and in your mouth. It was worse than coal dust, sickly. I hated the sweet smell – marshmallows.' Unless stone-dusters washed very carefully they awoke with their eyelids gummed together. Doug Ayres was never a stone-duster, but he remembers 'people walking around covered in black dust, but now and again a couple of men appearing looking like ghosts. Quite eerie.' The albinos among the blackbirds, miners said.

Stone-dusting on main roads and around the coalface was done at night.

Some non-mechanised pits drew coal on two shifts in the twenty-four-hour cycle, most, including all mechanised pits, on only one – the morning shift. The other two shifts, the afternoon (or back) and the night (or dead) were worked by a much smaller number of men, who carried out essential maintenance and got the pit ready for the morning shift. Mornings were most popular with Bevin Boys – they left afternoons and evening free. The main attraction of afternoons was that (until nationalisation brought the five-day week), mornings included Saturdays – five-and-a-half shifts; there were no Saturday afternoons to work. Nights were generally unpopular.

'Like most I worked all the shifts at different times but I didn't like the nights. Too spooky,' says Harold Gibson, who was at the Huncoat colliery near Accrington.

It was so quiet you could hear the faint whistling of the air pressure, the constant dribbling of coal or stone from the roof and the creaking and groaning of the pit props, which had me on edge all the time. You didn't hear any of that when coal was being extracted.

The main job of the back shift was moving the cutting machine forward to make the next cut in the coalface and, behind it, moving up the troughs or pans and the conveyors. The face was cut on the dead shift. The other main jobs on that shift were ripping and packing.

Ripping was heavy duty work. Rippers extended the main roadway, opened new 'gates' off it to the face or, if a new face was being opened, drove a new road (a 'heading') towards it. The job also entailed dealing with any problems with the roof. Under the stress of millions of tons of strata, slowly, irresistibly, the roof lowered – closed as pitmen said – and the walls pressed in; even the floor could rise like a partially inflated balloon. Rippers and their assistants ensured that the roads had sufficient height and width for the tubs to pass. They removed steel girders that the earth's pressure, in Michael Edmonds' words, 'twisted and deformed into Gothic shapes' and which even drove them into the ground like fence posts. They tore down lowered sections of the roof, mostly with picks but sometimes with shot firing, erected new girders and laid additional track. When Roy Doorbar went to work with the rippers he realised how tough it was:

Men did a yard in a shift, two if they were lucky. It was hard work for me too, laying the rails, shovelling the brought-down stone into a tub – big effort, stone is twice as heavy as coal – then take it to pit bottom, up to the surface and over to the grinding shed to make into stone dust. And back down to start over.

Packing the space (the 'gob') left behind the coalface as the cutting machine moved forward was another night activity that many Bevin Boys helped do. It involved replacing the temporary pit props put up by the colliers as they went with piers of stone and rubble. A packer was a skilled man; an artist, even. Geoff Rosling took his turn helping in the gob:

> It was impossible to support the overhanging rock all along the face. Moreover it would have 'thrown a squeeze', as they said, on the coal. Without the packs there'd have been a sudden fall – the odd prop would go with a loud crack when you were working, which was unnerving. It was remarkable to see what happened in later months. The earth gradually settled to reclaim the empty space. In time the packs were crushed to a small mass.

A number of men became engaged in jobs they'd been doing before becoming Bevin Boys: fitters who travelled around the pit repairing and adjusting machinery, electricians, surveyors. Ken Da Costa from Cricklewood, an electrician in a clock factory until call-up, was put into the electricians' shop at Arkwright colliery near Chesterfield. 'An electrician was considered an official, a gaffer – I had to wear overalls,' he remarks. 'Dead embarrassing at times. I was a Bevin Boy, the only one in the pit. You can imagine some of the things the miners said to me.' Peter Allen, with civilian surveying experience, was put on the surveying team at Ryhope. His job was to come in at midnight, determine with a theodolite the position to which the coalface conveyor was to be moved, and paint the new line on the roof of the seam. Another of his tasks was to inspect the old workings, looking for cave-ins, gas concentrations and the accumulation of water.

> We always took a safety lamp to test for gas and it was often unacceptably high. We then opened certain vent doors to clear it. Water was a problem with old workings and pumps had to be brought in. It could be quite frightening in the low returns. Most were in a terrible condition, with countless roof falls round which you had to crawl. Sometimes I wondered if I'd get lost and be condemned to wander around for ever.

Some men were just lucky in the jobs that came their way, at least some of the time. Geoff Baker's very first job at Norton colliery near Stoke was as assistant to the surveyors – he thinks it came his way because the manager had found out he played cricket and wanted him to play for the colliery.

In this cushy number, armed with a measuring tape, a ball of string and a lump of chalk, I determined the position of the roof supports. I strutted about the coalface in overalls and the miners, stripped to the waist and sweating profusely, viewed me with some suspicion. And I made matters worse by shouting at frequent intervals an expression I'd learnt, 'Come on, chaps, bollock it on' – which meant shovel more coal on the conveyor belt.

Arthur Gilbert's luck changed for a second time when he left the loader end and became his pit's safety officer. 'I was coming down the gantry from the pithead one day and the safety officer stopped me. "Grammar school?" "Yes, sir, Uttoxeter." "School certificate?" "Yes, sir." I became his assistant and shortly after he left and I got his job.' In the course of a month Gilbert examined every inch underground, taking dust samples for combustibility, using an anemometer to monitor the airflow, and pumping air samples into balloons to be checked for gas in the colliery laboratory. In the return roads, he felt, like Allen, that 'sometimes I was tempting providence', but nonetheless reckoned he'd fallen on his feet – even if he was frequently crawling about on his knees: 'A wonderful job, nine to five, down the pit around 11a.m. back up 2.30 – and even an office underground where I could comfortably take my leisure after my scheduled tasks.'

Other Bevin Boys made their own luck. Donald Whittle at in the Swinton pit near Manchester became friendly

with Alf, who was in charge of the signalling system. On the strength of my very sparse knowledge of electricity from school certificate physics, I persuaded him to put in a word with the undermanager so I could help him. I was thrilled to get the nod.

All the movement of tubs underground were controlled by a primitive signal and telephone system running on wet cell batteries and basic wiring. In the humid atmosphere, batteries needed to be topped up with distilled water, and wires got broken in roof falls. The job of the bell boys was to rush to any bell or phone out of order and sort out the problem. Rush is an odd word since walking a mile or more up a one in three incline with four feet of headroom isn't speedy. Anyway we looked after four coalfaces for two years. I lost the job and became the powder monkey when they employed real electricians, but for two years I thoroughly enjoyed myself. And I became very familiar with the runners and riders in the racing calendar. Alf doubled as a bookie's runner – he'd visit the colliers during the shift and would be up the pit promptly at two to place the afternoon bets.

At his Kent colliery, Ian McInnes talked his way off haulage on to the fitters' gang, 'which made me the bag carrier for a one-time collier, Tibby Hibbert, and we covered pipes, pumps, haulages and conveyers mainly in the north west district.'

There was a lot of water to be handled and pumped out of the pit, up to 20 tons pumped out for every ton of coal raised, so we were kept busy. In the face areas spontaneous combustion was taking place in the small coal left behind in the waste once the face had moved forward. This gave rise to the production of sulphuric acid,[19] which then appeared in the face water in a dilute form, ie six parts per thousand, and the main casing of the centrifugal pumps was progressively eaten away until it burst and had to be replaced. The acidic water also played havoc with the main pumping system, which led to local flooding in the roadways, and so we had to strip off and swim across to get to the pump site.

Tibby and I also installed eight-inch ventilation pipes in return air roadways, balancing them on our hard hats while we lashed them to the crown of the arches. This balancing act took my collar size from 14 to over 16 in a few months. The drawback was, in that position, starkers, standing in a tub, everyone would give my willie a ding as they passed, and I couldn't do a thing about it.

Water and McInnes's Bevin Boy days seem inextricably entwined. Allocated his own pit district in 1945, he introduced water spraying on the coalface in implementation of the Pneumoconiosis Act of the previous year, which tightened the rules on dust suppression. Soon after, he 'had his lamp stopped' – a summons to see the manager, who told him to use less water.

I was supposed to be spraying with a fine mist, but I'd overdone it. I was killing the dust beautifully, but the coal was coming up very wet. Most coal cleaning on the surface was wet – coal and water together, shake up, coal at the bottom and the dust suspended. But at Snowdown the system was dry – the shaking sent the dust in one direction and the coal in the other. McInnes completely bitched up the process.

*   *   *

Some colliery managers worried about Bevin Boys who wore glasses – something unknown among regular miners and a topic of wry amusement. There were too many to keep from going underground, but in some pits glasses-wearers were found labouring jobs on the surface; others were issued with safety glasses, Gerald Carey at Warsop Main colliery near Mansfield and Derrick Warren among them. Warren was sent into Newcastle when he was in training at Easington colliery near Sunderland 'and was provided with steel-framed specs with shatterproof lenses, the same material, I was told, as Spitfire pilots had on their canopies'. Jim Bates was promised safety glasses, but 'three and a half years later I was still waiting for them'.

There were other Bevin Boys who caused managers concern: those so physically weak or ill-coordinated they were either a hindrance or a danger to themselves and others. Under wartime regulations such men couldn't be got rid of. The answer, where possible, was to make them assistant time-keepers or give them the job of helping the checkweigh man. 'Help' is a euphemism. Checkweigh men had been known to cook the books in favour of the colliery and were so widely distrusted by the union that it paid for someone to work alongside them to keep a separate tally of tub weights – a check on the checkweigh man. A manager required union agreement to appoint a Bevin Boy.

That Bevin Boys went to be employed 'in or about a coal mine' was something of a catch-all. Generally speaking it was taken to mean that they had to be engaged directly in the production of coal, and becoming an assistant time-keeper or duplicate checkweigh man appears to have officially met the definition. Whether bringing a man out of the pit to work in the wages office met the definition is a moot point. Peter Archer, marking time before call-up as clerk to the Birmingham district auditor in Sandwell, did the job for the last nine months of his service at the Old Coppice on Cannock Chase in Staffordshire. 'They were in a muddle over wages and they needed somebody in the office, but it was against the rules, of course,' he says.

David Reekie's Yorkshire colliery certainly believed employing him in the office didn't conform to the regulations, but no one could calculate the wages or sort out the weighbridge documentation when the manager went down with severe bronchitis, and into the office he went. However, as Ministries of Labour and of Fuel regularly visited to inspect, 'whenever a stranger was seen approaching the site, Walter the banksman sounded a klaxon as a warning' and Reekie would 'grab my safety helmet, dive out of my little wooden hut, and disappear into the drift entrance like a rabbit scurrying down his burrow'. To make his escape less obvious, the colliery handymen were told to turn the hut round 180 degrees so that the door was on the blind side – no small task as the telephone and electricity supply and the connections to the weighbridge had to be disconnected.

Alan Brailsford went to work in another Yorkshire pit's office, coming to management's attention in individual fashion: irritated by the pit's inefficiency, he demanded to see the company managing director.

During my time the mine moved to some mechanisation and used compressed air to take tubs up inclines, which was fine going up, but down, they invariably came off the rails. I became so frustrated I told the deputy I wanted to see the MD. 'That's not bludy done,' he said. 'Well arrange it,' I said. I was invited to see the chap at head office at Nether Edge on the south of Sheffield. He spared me nearly an hour – a most interesting conversation. But as I recall the only result of my interview was to be invited to a cocktail party of the mine officials. I suppose I was an unusual individual.

Shortly after this, he relates, 'The deputy buttonholed me. "You a bludy accountant? Manager in wages has had heart attack."' Brailsford filled in for quite a while.

> Every week the miners wanted a sub, because they'd had a sub the week before. Some asked for a spare wage packet so that their wife wouldn't be aware of their real earnings. Some wives, who obviously knew what their husbands were up to or to prevent them spending the lot on beer, came in to collect their wage packet direct – a common thing at other pits, I believe. I was very interested in seeing the miners close up.

As a post office clerk, John Potts, underground at Shipley Woodside Piper colliery at Ilkiston in Derbyshire, had to put up with a lot of jokes about licking stamps, but turned that to his advantage,

> making sure the manager knew I had clerical experience – I was the man if he needed somebody. For a year I had a marvellous fiddle: down the pit at 6am to book the men in and do a spot of clerical work for the deputies, then back to the surface at 8, shower, and into the wages office, nice and clean.

The downside was that Potts saw a few miners too close up for comfort.

> Pay day had its moments. There was a system of recording in the lamphouse what shifts men had done. Some did double or even treble shifts and there was often confusion. Miners could get quite belligerent, though it wasn't my fault. Once, one pinned me against the wall. Most miners paid the blacksmith to sharpen their pick heads but some took theirs home to do themselves. This fellow had a blade with him and held it to my throat.

The 'in or about' regulation created a topsy-turvy world for some Bevin Boys. Bert McBain-Lee, who fetched up at Treeton colliery near Sheffield, relates the tale of Fred,

> who had a degree in chemistry and was asked to help out in the lab where the staff were experimenting with waste coal products to produce what eventually were known as coal briquettes. Naturally Fred said yes. Two days later a Ministry man called at the pit and the manager, who was very proud of his laboratory's contribution to the war effort, showed him round. When he came to Fred for some reason he asked him how long he'd been working on the project. And when he said two days, the Ministry man ordered poor Fred back down below immediately.

Four

Notes from the Underground

The coal-drawing shift. Bevin Boys, like their fellow miners, arrive at the colliery. If there's time to spare, some head for the canteen and a quick roll and a pint of tea. If, that is, there is a canteen: most pits still don't have them. Then the pithead baths to change. If, that is, there are pithead baths: three in four miners still come to work in their pit clothes. Then the lamphouse. Every man exchanges his token for his lamp. Up the gantry or the steps to the bank. A random search by the banksman for cigarettes, matches, anything of ferrous metal – potential dangers underground. The cage loads. Drops.

The pit is poised as men disperse to their work stations. And then it comes alive. And the rhythm of the shift establishes itself.

That rhythm could be disrupted by any number of eventualities: runaway tubs blocking a roadway, torn-up track, a break in the haulage rope, a conveyor grinding to a halt choked with coal, even a sudden roof fall. When something happened that gave men unexpected downtime the astute were ready for it. John Wiffen 'kept a bundle of *Listeners* hidden on the top of a girder at the highest part of the roof for such moments'. On 'many days' Meir Weiss 'managed to finish reading the *Manchester Guardian*'; he also practised his Hebrew on the side of the tubs, which another *kindertransport* Bevin Boy working at pit bottom corrected once the system was running again. Brian Folkes and the Bevin Boy gang on his junction had a piece of sheet metal on which to play chequers, using 'spare, at least I hope they were spare, nuts for the engine. The bush telegraph gave ample warning of the approach of the overman.'

The only official break in the shift was for the midday meal (usually around 11a.m.). Generally it could be relied on in mechanised pits, where the machinery was shut down; in pits where coal was hand got, some colliers – on piecework – didn't stop, and nor could their unlucky putters, whatever the regulations.

Food taken down the pit, and the break to eat it in, was most widely known as 'snap', though in northern parts it was 'bait'. Able to live at home in Blackburn and catch the same local train to Huncoat colliery as a number of regular miners,

Harold Gibson, on his first day, 'like the new boy at school, with shiny helmet and shiny toecaps', met 'some old hands on the platform who wanted to know all about me and had I brought my bait. This puzzled me and I said I was going mining, not fishing. After they stopped laughing they told me bait was my midday dinner.' He never forgot that first day for another reason: 'When it came to bait time imagine my embarrassment when, on opening my tin, I found that my mother had wrapped my jam butties in a nice white linen serviette, when the other men had theirs in yesterday's *Daily Mirror*.'

Almost all men, regulars and Bevin Boys alike, carried their food in a purpose-made tin that developed from the pair of mess tins issued in the army and which retained the same elongated handle, to hook it to the belt. What was the bottom of the top mess tin as it fitted into the other had become a sealed lid, which kept the snap (or bait) from going stale in the humid pit air – and which also kept out the rats or mice, even if it didn't always keep out the ponies.

Pits teemed with either one or the other; miners joked about low roadways by saying they were so low that the rats, or mice, were bow-legged. The first time Alan Brailsford went to work he had his sandwiches in a pocket of his jacket, which he hung on a girder, 'and the rats not only ate my sandwiches, they ate my pocket.' On his first snap break in the Deep Duffryn pit near Pontypridd, Dennis Faulkner expressed disbelief that there could be rats so far underground. At that, 'one miner told everyone to switch off their lamps. He then placed his lamp, still lit, a yard away and threw a piece of bread to it. Within seconds several large rats descended upon it.' Geoff Darby never doubted there were rats, as he'd learnt in training. 'I never forgot that we were told to do nothing if we saw a large number – there could be 200 of them on the move from one face to another and they could attack,' he says. 'I never did see an army of rats, but I was forever imaging there was one about.'

More pits had mice than rats. Hundreds lived in the old packs, particularly at junctions where Bevin Boys gathered for snap, and were sometimes trapped inside by 'the weight coming on': the inexorable settling down of the earth; their piteous squeaking could go on for hours. No one liked the rats; Bevin Boys hung pieces of bread on string and waited for them to appear, a lump of stone or coal at the ready. The mice, however, so small 'they were like big bumblebees', according to Jim Ribbans, were rather a source of diversion; it was common for them to run up and down men's arms and legs when they were sitting down at snap time, looking for crumbs.

Snap typically consisted of two sandwiches, a rock or lardy cake, and perhaps an apple; occasionally regular miners brought in tomatoes or onions from their garden or allotment, which they shared. Bevin Boys in lodgings had their snap made up by their landlady as part of their weekly rent; those in hostels ordered theirs there or at the pit canteen for a daily sum of 5d. The weekly 2oz ration of butter didn't stretch to daily sandwiches, which were usually filled with jam,

Spam, corned beef, dripping (termed Yorkshire marmalade in Yorkshire) or cheese; the cheese ration, like butter, was also 2oz, but like those in other heavy industries who carried their food with them, and agricultural workers, miners – and Bevin Boys – had an extra 10.[1]

'I didn't eat cheese, like my mother, so I lived on jam butties every day for three years,' says Harold Gibson.

> My father, a policeman, and younger brother enjoyed my extra ration. But when my brother went into the air force he sent home tins of jam. He was stationed with the Americans in the Middle East and I've no idea how he fixed that the stuff got to us but he did and it was full of real strawberries. I used to show blokes down the pit.

Alan Lane, at the Lady Windsor near Pontypridd, lived on a combination of cheese *and* jam: 'I had digs in the village of Ynysybwl and my landlady used to give me cheese in two pieces of dry bread and on top of that one piece of bread with jam. By the time I got to eat them everything was jumbled up.' David Reekie had a very different kind of landlady: 'She favoured me with the brown bits from the dripping but swore me to secrecy not to tell her miner husband.' Geoff Baker's landlady made him beetroot sandwiches because he asked her: 'They were nice and moist – and the beetroot stains helped cover up the black fingermarks.' No washing facilities existed underground; only the fastidious bothered to spit on their hands.

Young men doing tough physical work were understandably ravenous by snap time; Arthur Gilbert once opened his tin upside-down and dropped his sandwiches in the dust. 'I picked them up, dusted them with filthy hands, and ate them. It was that or go hungry. Howls of merriment all round.' In truth, grateful as most Bevin Boys were to get something to eat, all of them were more grateful for the sit-down – even naked. 'You'd sit on a broken rock to have your snap and you didn't notice,' says Ian McInnes. 'Working naked, the skin on your buttocks soon got hardened.' Sitting in groups, men with cap lamps dimmed them so as not to dazzle others; lighthouses were dimmed by covering them with safety helmets. Anyone in Yorkshire who forgot was likely to be asked: 'Doest tha' want to take a photograph, lad?' At the lighted but often cold pit bottom, men sat with their lamp between their thighs for warmth.

Snap time lasted only fifteen or twenty minutes but humidity, coupled frequently with poor air quality, and suddenly relaxing after hard graft, made the urge to fall asleep almost impossible to resist: a punishable offence underground. 'Our underground manager was a shortish man with a mole-like face and a mean temperament who delighted in catching men dossing after finishing their quota of work or extending their snap period,' says Peter Rainbow, who was down the Leycett pit near Stoke.

He'd approach in the dark with only a lighthouse lamp, so you'd think it was just another miner, then when he was a few feet away he'd turn on his dazzling 'shinie'. If he caught anyone, asleep he'd pour water over their faces. He put the fear of God into the more timid among us with his threats and strong language.

'Most of the time I was nearly exhausted by snap,' says Ken Sadler. 'Sometimes I was so tired that I didn't bother eating but slept until called back.' Ian McInnes once fell asleep when he'd eaten his snap having spent an hour mending a pump 'lying in a foot of water with my other half exposed to a howling gale'. He'd retreated to a warm main and tail engine house. 'The next thing I was awakened by a bull's-eye spot lamp shining in my face. It was the mine manager doing an inspection in the company of the union pit rep.' He was fined 5s. Norman Bickell was another who fell asleep, but he was awoken in more startling fashion:

I became aware of the most curious feeling at the top of my thigh, as though something was vibrating in my trousers. I clapped my hand on my thigh and the vibrating stopped. Dropping my trousers I found the carcass of a crushed mouse not much bigger than my thumb. From then on I always tied string around my ankles.

The call back to work came all too soon for everyone, those who'd nodded off jolted awake by whistle or bell or just hammer against steel and the cry in these words or others like them: 'Up, up, we're drawing coal again!'

*   *   *

Perhaps only at the start of the Monday morning shift was it possible to distinguish the Bevin Boys from the miners. Worn and patched as the miners' pit clothes were, they'd been washed; the Bevin Boys' were as filthy as they'd been at the end of the previous shift. Miners (their wives, anyway) thought it mattered. The small minority of Bevin Boys who lived at home were spruced up to start the week (their mothers, including Harold Gibson's, thought it mattered); the rest didn't see the point. If they could be bothered to wash a shirt (or have it washed for them) and darn a sock, that was it.

After training, recruits had to hand in the overalls with which they'd been issued. A few bought others. Seeing these, the miners laughed, as they did at Alan Brailsford for 'turning up to work in a rather good suit', for which he was 'unmercifully teased: "Who thee think t'ar, bludy managing director?"'

Down the pit men wore anything: a jacket or battledress top was useful for stowing away bits and pieces like the lamphouse check, chewing tobacco or snuff or, in naked-flame pits, spare carbide for the lamps and even, where smoking was allowed, a cigarette tin. Like the miners Bevin Boys slit 'gamekeeper's pockets' in the lining for snap tin and drink container, to leave the hands free; in some pits Bevin Boys copied the fashion for waistcoats, sewing big pockets to them for the same purpose.

Trousers were the problem down the pit – they were soon ripped to shreds. 'You were always catching some part of you on damaged metal tubs or splintered pit props,' says Geoffrey Mockford. 'There'd be a loud ripping sound and you'd find a flap of material hanging down. That got in the way and was a potential danger too.' Old lighthouse lamps occasionally leaked acid and, as Mockford points out, as those on haulage hung theirs from their belt 'to get a good view of the haulage clips', not only

> were legs made sore, but your trousers were turned into a type of lacework right in the area where they were most needed. This was no great concern below ground, but we did have to walk from the pit top to the baths which, to say the least, was embarrassing.

To do running repairs, some Bevin Boys brought down needle and thread. When his knees were out, John Wiffen used to bring down patches already cut from some coarse material and

> if there was a long pause in work for any reason, I'd take off my trousers, sit on the bench, set the two knee patches from the inside of the leg with a single thread over-and-under stitch to place them, then bind them in with a double thread over-and-over stitch. The patches often outlasted the trousers.

A Bevin Boy at the Yorkshire Main pit was evidently in Wiffen's class when it came to needlework: Les Thomas was so impressed with his patches he asked him where he'd learnt. 'The answer was sewing mailbags – he was one of those who'd initially refused to be a Bevin Boy and been to prison with hard labour.'

But needles were of ferrous metal and theoretically could trigger an explosion underground, and while that risk was infinitesimal many pits forbade them being taken down. Jim Bates didn't know that and found out 'when the banksman frisked me for matches and thrust his hand into my pocket with my needle in it. I learnt a lot of alternative King's English in a very short time'. Thereafter he stitched himself together with shot wire, as almost all Bevin Boys did. There was always plenty of the stuff around in the pit but, Geoffrey Mockford says, 'you had to be careful that a piece didn't have a live detonator attached to it. Sometimes detonators failed to go off and were supposed to go to the surface with the coal – but they were often left lying around'.

Bevin Boys wore their pit clothes until they'd all but dropped off before finding something to replace them. Nobody bought anything if they could help it: the government had given Bevin Boys a generous allowance of thirty extra clothing coupons but not the money to put them to much use other than for the odd pair or socks or a towel or two. In dire necessity Bevin Boys went to jumble sales or places that sold ex-WD uniforms. Derek Agnew saw one Bevin Boy in his pit wearing an Italian PoW chocolate-coloured battledress. Many an ill-fitting garment was dug out of the back of fathers' wardrobes. Quite a few old football shirts were pressed into service. John Wiffen became friends with a postman who gave him some cast-offs. Les Wilson's father gave him the trousers of the evening suit he'd worn when he was the drummer in a group that played in the cinema during the silent film era, and Harold Gibson's policeman father acquired him a full uniform, 'without the chromium-plated buttons, but I was still the smartest bloke down the pit. Good solid serge trousers – protection against jagged edges.'

'Frankly,' says Geoffrey Mockford, 'we looked like rag and bone men.' Norman Brickell cheerfully admits to probably looking worse. 'I was dressed in virtually rags that never got washed. I must have looked a pitiful wretch. It's easy to see how I came by my nickname, "Shabby", later abbreviated to Shab.' A proportion of miners who travelled to and from the colliery in their pit clothes had a predilection for colourful scarves or mufflers that brightened their drab apparel and some Bevin Boys followed their example.[2] The dirtier and shabbier they were, the more bizarre their appearance could be. As one Bevin Boy observed: 'We looked as if we were off to a tramp's ball.'

Many Bevin Boys' pits boots were in the same condition as their clothes. It was sensible to clean and grease boots liberally – more modern pits had an electrical device with revolving brushes and a grease applicator on the dirty side of the pithead baths (see page 87); all at least provided long-handled brushes and a tub of grease. But many Bevin Boys left their boots just as they came up from underground. On average boots gave reasonable service for eighteen months, a year if men worked constantly in water. Boots that weren't looked after lasted for much less – and cost up to 30s to replace. Bevin Boys resented the expenditure and very few made it.

'However you looked at it taking care of your boots made sense,' says Geoffrey Mockford. 'Greasing boots kept them pliable and preserved them. Left ungreased in heated lockers they became hard and they'd crack when you put them on. Many Bevin Boys found their boots impossible to wear after a period of absence. Some were told off for wearing boots in a terrible state.'

Whether they cared for their boots or not, a large number of Bevin Boys continued to dislike wearing them. Norman Brickell replaced his with a pair of American GI army issue ('Marvellous, fantastic, the most comfortable things I've ever worn'); others switched to wearing clogs, which many miners, particularly in Yorkshire, Lancashire and Durham, continued to prefer.[3]

'Boots: horrible, inflexible, clumsy things,' retorts David Reekie. 'I wasn't able to keep up with a speeding tub travelling on a down grade. The accepted and only way not to part company was to slide on the rails. And for that you had to have clogs.'

Clogs cost about a pound, but Bevin Boys discovered that for about 3s 6d they could buy a pair of soles at a local cloggers together with a set of four irons, the two smaller ones fitting inside the others on the centre of the soles. These not only protected the wood – but allowed hauliers to put a foot on each rail.

Clogs lasted only about three months before the sole wore right through, but Bevin Boys gladly paid out the 2s it cost to replace them (they reused the irons). 'Clogs were perfect for tramming [putting],' says Reg Taylor.

> But they also kept your feet warm in winter, particularly in snow – there was nearly an inch of wood between your foot and the ground. And they self-customised. What happened was that they tended to fill with the coal dust and ground-up stone dust, and with the sweat from your feet this soon formed a sort of concrete, which adapted to the unique shape of your foot and became oddly comfortable.

Living in areas where mining blended with textiles, Bevin Boys followed the custom of embellishing the toes and heels of their clogs with pieces of carding leather. 'Everybody knew somebody who worked at t'mill and so there was usually a supply of used carding leather, which was used on rollers for drawing yarn,' Taylor says. 'The corrugations did give a stylish appearance!'

*   *   *

Few men would disagree with George Ralston, who served his time at the Lady Victoria, Newtowngrange, Midlothian, that 'water was more precious that food down the pit'.

How much water a man carried down with him largely depended on the depth and humidity of his pit. Doug Ayres remembers that men who worked in or near the face in his Staffordshire pit, 'who'd taken in six- or eight-pint containers and drunk and sweated the lot would beg for a mouthful of water from the haulage lads at the end of a shift'. Six or eight pints were as nothing in Jim Ribbans' Kentish pit where 'anything up to 24 pints wasn't out of the ordinary – 16 was quite average'. Some pits circulated a water wagon, a caulked tub with a sliding hatch to stop the water slopping out over the rough ride and collecting dust along the way. In Durham, Dennis Fisher's pit had no need of one: 'We were under a lake and we refilled our drinking jacks from what dripped from the roof. Pure and sweet – nothing like it, mun.' Some regular miners declined to fill their containers from the taps on the surface: they brought boiled water from home in the belief that it prevented them from getting a chill on the stomach. Some added a pinch of salt to replace what they lost by sweating and had their imitators among the Bevin Boys.

Relatively few men preferred tea to water for their fluid intake, which they brought down cold, without milk: milk curdled underground. Cold tea was an acquired taste that was quickly acquired. 'A large bottle of cold tea,' says Alan Brailsford, 'was like champagne.' There were, however, always those who wanted their tea hot, and if they worked at pit bottom they were able to get it at snap from an enterprising banksman for a shilling a week. Anywhere in a non-gas carbide-lamp pit men could get a piping hot brew. 'What we used to do was take some crystals out of the lamp, pour a bit of water on them and light them,' says Ken Sadler from his days at Blagdon drift, not far from Seaton Burn in County Durham. 'You got a nice little fire going in no time.'

Canteens sold drinking jacks made of tin or aluminium in sizes up to eight pints, but many men used lemonade bottles. This wasn't against any regulation but, given the many circumstances underground in which they could be broken, seems to have been a dangerous practice. Reg Taylor used a bottle for his tea and one once fell through a hole in his pocket, shattering into pieces on a rail.

The miners told me off for being careless, not for having a bottle – most of them had bottles. We were never told we couldn't bring down glass; even at the training centre at Pontefract some people had bottles: one Bevin Boy tried to kill a mouse there and smashed his to smithereens. As I remember, the miners occasionally broke bottles in the pit. The thing to do was to gather up the pieces carefully and throw them in the gob.

Miners had several ways of taking the edge of their thirst to ration their precious water supply. 'A couple of hours in, you'd be thirsty as hell,' says Roy Doorbar.

But you only had the one bottle, so to save it you'd pick up a small piece of coal, spit on it, wipe it on your shirt, and stick it in your mouth. Sucking it, moistened it. You took a gulp of water now and again, but you needed to keep some for the end of the shift six hours away.

In the Yorkshire Prince of Wales, Phil Yates and other Bevin Boys took to sucking liquorish root.

There was a liquorice factory next to the Prince. Nowadays all the liquorice comes from Iran and Turkey, but then they grew it in the fields and the smell permeated Pontefract. Many of the miners' daughters worked in the factory and miners brought down the root to suck. It was certainly better than a chaw of baccy.

Some miners took snuff to clear their nostrils and a few Bevin Boys took to the habit; most miners chewed tobacco twist to moisten their mouth and many

Bevin Boys tried it – once. Harold Gibson's experience of doing so was typical: 'I was warned not to swallow but I did. I was so sick I thought my end had come.' Arthur Gilbert swallowed his first time, too, and 'in a few moments my senses were reeling'. He, however, persisted and acquired the taste.

> It was almost impossible not to swallow sometimes, but I got used to it – a great remedy for constipation. And I became a master spitter: compress the cheeks and lips, squirt, and the juice would travel a considerable distance. I got so good, above ground I could hit a fly.

*   *   *

During training at Askern Main near Doncaster the deputy asked David Reekie to accompany him into a side road with about 4ft headroom, told him 'to take a deep breath, if I were you, lad, and hold on to it as long as tha' can'; and with that advice pushed his way through curtains of overlapping brattice cloth. 'Just beyond the curtain the atmosphere changed from hot and unpleasant to stinking hot and vile,' Reekie remembers. 'There before us in the fluctuating beams from our headlamps was an Elsan [chemical toilet] of many years standing, brimming.' He barely heard the deputy say they were going to carry it out and empty it before he'd fled back through the curtains trying not to be sick.

There could hardly be plumbing in coalmines, but there were supposed to be toilets, as clauses 106–111 of the Coal Mines Act 1911 made clear:

> A sufficient supply of suitable sanitary conveniences shall be provided . . . below Ground at or near the Pit Bottom and at suitable positions along the main road. Every Sanitary Convenience below Ground shall have a portable receptacle constructed of metal and provided with a metal cover. Every Sanitary Convenience shall be kept in a cleanly and sanitary condition, and in good repair, and the receptacles of all conveniences below ground shall be emptied and cleaned not less frequently than once in every seven days and oftener if necessary. No person shall relieve his bowels . . . on any roadway below ground except in one of the conveniences provided in accordance with the foregoing regulations.

No colliery at the time was known to comply with these regulations; some did have a chemical toilet near pit bottom or more likely a bucket or two, for those with the stomach to avail themselves. That was all. Indeed, few if any collieries provided toilets on the surface, also a regulation of the 1911 Act.

While he was in training at Woodhouse near Sheffield, Derek Agnew, a schoolboy from Kingsbury in London, and his intake were told by their instructor: 'About hygiene I've only one thing to say. We "forgot" to build toilets in coal mines.

You know that, so to save acting like a lot of dogs, just get into regular habits.'[4] Some things, however, are beyond control: in the pit men were obliged to meet the most urgent of human needs as best they could, coating their excreta with stone and coal dust before dispersing it in the coal on the conveyors or in the tubs for conveyance to the surface – another reason why working on the screens was so unpleasant. Some regular miners brought down old newspaper to wipe themselves but Bevin Boys, like the majority, used pieces of coal or stone, or a handful of straw or anything else that came to hand.

Smells travelled in the airflow from one end of a pit to the other (if a man at pit bottom peeled an orange, a rare acquisition in the war, it could be smelt on the coalface): sweat and the lingering fumes of shot firing, the sour smell of fresh-hewed coal and stone flour, ammonia from the stables in pits with ponies, and their dung. But the worst of smells were human, a stench in the big pits with thousands of men underground, and it hung around all day. 'In the unforgiving airway,' comments Doug Ayres, 'there was nothing to be done except make colourful remarks, some of which still ring in my head.' Combined with all the other pit smells, it made numbers of men queasy. But not Victor Simons, who worked in Bolsover colliery in Derbyshire. 'I had been at boarding school and was not sensitive,' he says cryptically. Monday mornings weren't looked forward to by anybody. 'The miners were notorious at weekends for overindulgence of alcoholic beverages,' explains Arthur Gilbert. 'Not for nothing was the local brew referred to as "Parker's purge".' Commonly, miners downed twenty pints on a Friday night; in the miners' welfare club in Ryhope, according to Peter Allen, 'they gave a discount of a penny on every pint drunk over 13.' It didn't help that in many areas the miners' weekend treat was a plate of mushy peas.

Relieving the bowels below ground wasn't a private act; there were many disused tunnels and crevices but these were highly dangerous because pockets of deadly methane could collect in such places and had to be avoided. Men simply moved downwind, away from their workplace and other men. In such circumstances, Geoff Baker remembers, 'there were no niceties of privacy, and private parts were anything but private. I was amazed at the endowment of one miner, which was the largest I had ever seen or even thought possible.' To the collier he worked with Baker remarked that

'Sam is well blessed.' Joe looked at me and smiled. 'Sam and 'is faither and 'is brother 'ave got a yard betwain 'em,' he said. I looked at Joe in disbelieve. 'It's true, ah tell thee,' said Joe. As I never had a viewing of the other two I had to take Joe's word that it was a fact.

In pit parlance, for obvious reasons, the act of defecation was referred to as 'on the shovel'. The phrase perhaps had more resonance for Baker than most Bevin Boys, partly because of the reason he acquired a shovel in the first place – through losing his cushy number on the surveying team. The manager had spotted him making his way out before the signal for the end of the shift. Next morning he found a note on his lamp, which made it evident that cricketing prowess no longer stood him in good stead: 'Since you have seen fit to disregard the confidence and trust I put in you, you will start loading on the Cockshead seam as from this morning.' That meant he needed a shovel and a pick, and he went to the stores to get them. Like David Reekie a man with an ear for dialogue, Baker relates the encounter that followed:

'Wot's they want wi' a pick and shovel? Oose it fer?' the storeman asked. 'It's fer mey, o'course,' I replied, doing my best to put my case in the local dialect. 'Wot at they gooin' ter do wi' a shovel?' 'Well,' I answered, 'I've got to use it.' 'Well, whot sort er shovel dust want, wot weight?' I was rather stumped with this question. I was aware that there were different weights in cricket bats but not shovels. With some reluctance the bemused storeman handed over suitable implements and I turned to go. 'Ere theist fergot to sign fer 'em. Thee anna, free, thee knowst. Theyst gotta pee fer 'em ite a thee weeges. They knowst that, dusna?' 'I do now,' I said.

Set to work 'bollocking it on' himself rather than telling others to do it, Baker shovelled for some hours before he wanted to meet the call of nature. It was the first time he'd worked with Joe – 'Big Joe to everyone because he was a big man in every way, six feet four, 16 stone and big-hearted, a fantastic bloke' – and was embarrassed to make his need known.

Eventually I said, 'I want to be excused.' Big Joe didn't know what I meant. 'I want to go to the toilet.' 'A piss or a shit?' 'I want to sit down,' I said, lord knows why, it was a phrase my mother used. 'Tha's na gooin' ter walk o' that wey over a mile ter the pit bottom just ter 'ave a bloody shit,' exclaimed Big Joe who, by this time, was getting rather irate. 'But I haven't any paper,' I said, trying to make a feeble excuse. 'Yer dunna want any paper,' retorted Joe. 'Use them wood shavings.'

In a poem entitled 'Ode to Joe', Baker wrote these stanzas:

I suppose one of the things I found irksome
In my new cloistered life down the pit,
Was the question of personal toiletry,
Especially with nowhere to sit.

I sought some advice from my workmate:
'Where's the best place to go in this hovel?'

His answer was not what I wanted:
'Get thee pants down and go on thee shovel.'

*   *   *

Baker has another reason for not forgetting his first morning at the face. He turned up determined to show the regulars what he was made of. He got stuck in and in his estimation he

loaded more coal than any miners had ever done in the north Staffordshire coalfield. I never stopped, was sweating profusely, had pains in all parts of my body and was absolutely knackered. I started to slow down, thinking it must be near snapping time. 'What time is it?' I asked Joe.

It was about quarter to nine.

No one had watches underground. But men judged time by the number of tubs they filled or handled and Bevin Boys learnt to do it. Deryck Selby would never have had an idea of the time if he hadn't: when he asked the miners in the Empire pit in Cwmgwrach, in the Vale of Neath, they always replied in Welsh. 'They could speak English but I suppose it was their way of saying I was the stranger in their midst,' he says. 'To be fair, on the bus and among themselves down the pit, they spoke Welsh.'[5]

The end of a shift was known by different names in different parts of the country; 'loose it' was probably the most widely used term, but 'lillicock' seems to have been more common in the Midlands.[6] Again, different pits signalled knocking off in different ways, usually with five or seven rings that were flashed along the bell wires, or by blasts on a whistle that were picked up by other whistles at the districts inbye.

It was accepted practice in many pits that when a collier had finished his stint it was loose it for him, whatever the time: 'He could "Eff off home" – or "wom" in local dialect,' Alan Gregory says. He adds:

There was one exceptional character on the face I most worked on, Ernie, or Narny as he was called, Fletcher, who was always the first man to away. He was stocky and broad in the shoulders and chest, and he'd hammered his shovel to splay it out to take the maximum possible load. When he'd finished he'd come down the face on his knees in the shaker pans, head bent to dodge the straps and bars over them. I, as you might expect, was usually one of the last to come off.

No sooner than loose it sounded, then men were away towards pit bottom. Syd Walker was so eager to get there he often ran, against all rules, crouched in an awkward position, watching the roof for any pieces hanging down and the floor for any fallen rock. My problem was that I'm over six feet tall and while the 'drawing roads' were the highest places in the pit, they were mostly five feet. It could take you almost two hours to get out. There were times when I smacked against the roof and was sent flying on my back.

In his very wet pit in Kent, Brian Folkes sometimes had to make a choice at loose it. 'The water on the roadway outbye from my work station could get so bad I could either wade out chest deep or ride on the coal conveyor – a perilous operation with coal chutes from face conveyors giving about ten inches headroom.' The virtue of wading was that, by tradition, men who'd been working in very wet conditions were allowed to the head of the queue for the cages.

There were always men so impatient to get away – and a long walk from where they were working – that they sneaked forward before the signal, otherwise, especially in pits with a large workforce, they'd be at the back of a very long queue. Eddie, the tadger man David Day worked with in Staffordshire, with a bus to catch to his home 20 miles away, had a golden rule: be at pit bottom for lillicock;

> so each day we smuggled ourselves off the face before time, provided of course we were no longer required on the face . . . We were handicapped by our lamps which could be seen from a long distance in the dark but the odds were that we saw the gaffers' lamps first, particularly as they carried the unmistakable shinies and safeties.
>
> Several times Eddie and I lay curled up in a manhole, hardly daring to breath while the silver rays of the gaffers' lamps flashed up and down outside. On one occasion, when I weakly chose to give myself up rather than risk hiding, I gazed aghast as the overman stood threatening to report me for misconduct, at the same time spitting his tobacco juice into the tub in which Eddie was concealed.[7]

It was a matter of fine judgement for those creeping up to pit bottom from the coalface area (and scores of regular miners and Bevin Boys alike did) as to when to spring out of the darkness; too quickly and their black faces showed up among the (comparatively) white faces of those working there, the overman would spot them – and hold them back. The scramble for the cages at the end of a Saturday half-shift was particularly acute in football towns; buying a place in the queue for a shilling was common.

The walk out of the pit, after a shift worked, was a harder slog than the walk in; and in most pits the final stretch was up a long steep incline – 'lovely going downhill to the face,' comments Geoff Darby, 'but when you were tired it was a bugger'.

John Wiffen and some of his fellow Bevin Boys, who'd marched towards pit bottom 'singing scurrilous songs', refused to recognise this.

> You might have thought that we would stop singing when we climbed the last 600 yards of one in three. Not a bit of it. It was a point of honour to increase pace and sing all the louder until, approaching the top, our lungs screamed for rest and air and the blood in our legs seemed turned to water. We never let it appear that we were at the point of dropping. We were young and as fit and full of life as we should ever be.

Pit bottom workers might have had cleaner faces, but everyone came up in some degree of filth, most caked in coal dust but also often in stone flour, clay or grease, the majority 'with ears literally filled with dust, a nose that when you blew it had enough coal for a fire, and phlegm and mucus so thick with coal dust that you could never see any white in it,' says Les Wilson. 'I've since wondered if you needed any fuel at a miner's cremation with the amount of coal dust that must have settled in their lungs.' As dust and dirt dried on them, those who'd been working in water, like Harold Gibson, 'sometimes came up walking like Frankenstein's monster'. Doug Ayres on occasion came up covered in grease, loathing the grease wagon with a passion:

> This was full of thick black grease and circulated around the pit all the time for people to lubricate points and switches. When it came round it was difficult to get hold of the damned thing – we'd get pieces of paper sacks to cover the wagon sides, but that was never entirely effective and on occasion you'd get absolutely covered in gunge.

As a cage began its ascent, men in it automatically bent their knees as they did coming down, though now they closed their eyes: after hours of winding coal every surface of the cage was thick in dust. The engineman accelerated to the halfway point where the cages passed each other in the shaft, then braked – with the reverse sensation that braking gave on a descent: the cage felt as if it was no longer rising but falling. To Roland Garratt, 'that, if anything, made going up more frightening than going down'.

After a good shift, in Wales at least, the regulars would sometimes sing as the cage rose; 'Sospan Fach' and 'Bread of Heaven' were John Wiffen's favourites and it seemed to him that 'the unanimity with which they were sung to accompany our upward flight reflected a completeness not found elsewhere'. Most Bevin Boys weren't given to such introspection; they were just glad to reach the surface. Deryck Selby found it 'a joy to see the blue sky and hear the sparrows'. The late Ted Holloway, one of several Bevin Boys to make a reputation as a miners' painter, appreciated the daylight above all. A forestry worker from Fair Oak in Hampshire, used to being in the open, he at first found going down the Craghead pit near Durham unbearable, because it deprived him of daylight. 'There must be light somewhere,' he wrote home. 'Perhaps it's over Africa or Cheltenham.'

Harold Gibson's only thought was, 'I'm glad to be alive. For three years that feeling never left me, another shift over and still in one piece. I'm sure I wasn't the only one to say a silent prayer.'

*   *   *

Smokers were always the ones to catch the last cage going down for the shift. Coming up, when other men rushed to the lamphouse to hand in their lamps for recharging and get back their tally, then head for the showers in the collieries with pithead baths or hurried away for their buses or trains, smokers hung back for the first cigarette in eight hours.

Most miners smoked heavily; non-smokers like Deryck Selby, who in the mornings caught a bus already filled with smoke, 'had great difficulty breathing'. But the forties was an era in which nearly everyone smoked, including many Bevin Boys. As David Reekie observes: 'It was the accepted thing. An example of the North–South divide was the miners' choice of cigarette. The worker in London smoked Players Weights, but his Yorkshire counterpart stuck with Wills Woodbines. In those days we Southerners referred to Woodbines as coffin nails.'

When they first joined their collieries, Bevin Boys innocently handed in their cigarette packs to the lampman for safe keeping. They stopped doing so when they realised, like Jim Ribbans, that 'if I left too many there were sure to be one or two missing'. After that they put their cigarette in Oxo tins, which they hid in the dust of nooks and crannies between the lamphouse and the bank, just like the regular miners.

Desmond Edwards, a non-smoker, was interested in the psychology of the smokers around him. 'I can honestly say I can't recall any miner express an urge to smoke during a shift,' he says. 'The craving simply switched off. But on reaching the surface the craving reasserted itself, there was an almighty rush for hidden tins – and the sweet mountain air was replaced with acrid fumes.' Roy Doorbar didn't miss a cigarette while underground but arrived on the surface gasping. 'The first thing I would do,' he says, 'was find my half-smoked fag, the last one before the cage, and the matches hidden in the dust, spit the dust off and light up. Oh boy, the first drag on that fag was heaven.'

Men were able to smoke down some pits: 'naked flame' pits, a few deep like those in the Kent coalfield, most of them elsewhere shallow drifts, and all of them very wet – the water dissipated the gas. While some of these pits allowed men to smoke underground, others forbade the practice. At Chislet in Kent, Anthony Shaffer thought the ruling perverse; so did Ken Sadler at Blagdon Drift in County Durham where, he remembers, they were searched for cigarettes and matches.

You could be fined immediately for having them but most of us did, hidden about us. The overman often checked underground. He usually came through ventilation double doors and you knew he was coming as you heard the suction, which warned any of us who were smoking to douse our fags. He'd come through sniffing the air and asking if anyone had been smoking. No, nobody'd been smoking. What was farcical was that when he'd left on his rounds we'd open the vent door and find a small pile of ash where he'd tapped out his pipe.

Bevin Boys today are adamant that they never saw any miner in a mine where gas was a hazard having a cigarette – with the exception of Peter Rainbow. At one point in his service he was driving a new heading with an older miner who 'frightened the life out of me at snapping when he'd light up. He'd first test for gas, but he was taking an awful risk. I persuaded him to stop – I suspect he thought I'd shop him.'

*   *   *

Conscript Bevin Boys and ex-servicemen who passed through training centres with modern pithead baths expected all collieries to be similarly equipped. Most were to find out otherwise.

At the start of the war only seventeen of the country's 1,886 collieries had such amenities; by the end, another 345 did. Very little new building was allowed during the war, but coalmining was so important to the war effort that the government made pithead baths one of the exceptions. The money came not from the colliery companies but from the Miners' Welfare Fund.[8] All these baths were of a decent standard and followed the same layout: a dirty side into which men came from the pit, where they left their work clothes, helmets and boots in heated lockers or raised them on racks or in baskets into heated air at the ceiling; the shower area into which they went with their towel and soap; and the clean side where, after showering, they got dressed in the clothes they'd left there at the start of the shift. They carried their snap tin and drinking container through from dirty side to clean side.

There were pithead baths at some other collieries paid for with money raised by local miners' institutes or clubs, but these were inferior and in some places basic in the extreme. Hem Heath in Staffordshire, according to Jim Bates, 'had pithead baths of a sort – an old Nissen hut'. At the small (200-man) Huncoat colliery in the Midlands, Harold Gibson remembers the shower room being about 20ft by 10ft,

with showerheads coming out of the bare concrete walls, no cubicles like at Swinton in training, no glazed tiles. It paid to get your shower over quickly, too. If you dallied or were the last in, the bath attendant would hurry you along by hosing you down with cold water. And there was none of your clean side, dirty side, or heat drying either. We had the one locker. Huncoat was a wet pit, so if your clothes were wet, they stayed wet. And you had to be careful how you put your helmet into the locker when it and the rest of your gear was wet. If you shoved it in and it wasn't lying flat, it became misshapen – helmets were strong but they were only composition, made of wood, jute and rag pulp, and if you ruined your helmet you had to pay for a replacement: two and six to three shillings. Your clean clothes weren't as clean when you put them on as when you put them in the locker! At Swinton I used to come out like a new pin, not at Huncoat. Still, it was better than travelling home in my pit clothes, which I had to do for months – I had to wait for a locker.

At Chislet in Kent, men's pit clothes were hoisted to the ceiling on the same hook that took their clean ones, inevitably transferring dust; worse, with men working different shifts, clean clothes hung alongside and brushed against pit clothes. And, Derek Agnew noted, 'one cannot help rubbing against dirty bodies and clothes.'

> I had just finished showering when a dirt-encrusted collier came into the same cubicle and splashed most of his dirt on to me. After cleaning myself again I gingerly walked over to my clothes only to find that en route my feet had picked up most of the coal dust on the floor . . . After dressing I got my hands filthy again in hanging up my dirty clothes and handling the grease-soaked pulley rope.[9]

Men didn't hang about on the dirty side in clean/dirty baths. 'There was always an unwelcoming aroma of stale sweat in the heat,' says Arthur Gilbert. 'Some of the occupants weren't too particular about hygiene and seldom took home anything to be laundered. By next shift the socks would easily stand by themselves and everything was stiff as cardboard.' In John Wiffen's words, 'the too-human smells, particularly of long-treasured boots, made exit from this side speedy.' Those who stayed, enjoying the greenhouse temperature were the cockroaches; during the summer break the baths were fumigated, if not oftener.

It wasn't only the Welsh who had the urge to sing in the shower area; Yorkshire miners at Welbeck colliery, which had several hundred cubicles and where, according to Stan Payne, 'everything possible was tiled and the acoustics were magnificent, someone would start to sing and others would join in, frequently with harmony. It really was some sound, especially around Christmas with the carols. Just like some cathedral.'

Almost all Bevin Boys remember the soap – a scarce and rationed commodity during the war.[10] Some collieries made miners pay for it ('Ten pence for a bar from a counter near the entrance,' says Geoffrey Mockford, 'but at least no coupons'); issues of carbolic seem to have been rare (Doug Ayres remembers 'big, hard yellow blocks of laundry soap with the word Bibby on them') but most got good, scented bath soap, even in Harold Gibson's bathhouse. 'Lovely soft white Windsor soap, so soft you could squeeze it,' he says. Stan Payne had 'beautiful, mild white soap, a big bar per week and I do mean big: with care you could make one last so that every five or six weeks you had a spare. I divided my spares between my landlady and my mum.'

It was traditional practice in pithead baths for men to scrub their neighbour's back, which he couldn't properly reach himself. 'You washed somebody else's back and he in turn would wash yours,' explains Doug Ayres. 'We tried to avoid Len who was massive, maybe six foot six, very muscular, probably a 60-inch chest and very, very hairy. You really had to be on piecework to scrub his back, it was like washing a gorilla.' To the amusement of many Bevin Boys, some older miners were shy – they wore bathing trunks in the showers. A miner called Fred amused Stan Payne more:

Fred was presumably bald but was never seen without his cap. Coming up, he'd stick his head, with helmet, into his locker and re-emerge with his cap. He went into the shower cubicle complete with cap, drew his curtain – which nobody did – and re-emerged with cap in place. Needless to say, poor old Fred never got his back washed.

For most Bevin Boys their shower, with plentiful hot water (coal-heated on site), was the best thing in coalmining. For Jim Bates, getting into clean clothes was 'a daily delight', just as 'climbing into your pit muck' was a daily disgust. Some Bevin Boys who lived in hostels used the pithead showers before going back and even showered again when they got there. But no one was every really clean; as David Day remarks, 'your underwear was black as the coal worked out of the pores'.[11] And unless they applied Vaseline with a cloth to their eyelashes and the soft tissue around their eyes, everyone, as Doug Ayres put it, 'left looking like Elizabeth Taylor'.

*   *   *

When war broke out, coalmining had more canteens than pithead baths, if not many more. By 1942 there were fifty-nine; by 1945, 912, full hot meals provided in 566. Some pits with canteens even ran to what would now be called a snack bar, too. Again, the money (£2.7 million) came from the Welfare Fund.

Bevin Boys at collieries with canteens counted themselves lucky and patronised them with gusto. Meals were off-ration and subsidised and were deemed even better value than the British Restaurants. For a shilling there was typically roast meat or steak and kidney or shepherd's pie, served with potatoes and two veg – second helpings allowed – followed by steamed puddings or apple pie with custard. When Doug Ayres came off night shift, he'd

> sometimes expend a little of my sparse wages on a plate of porridge and golden syrup. Alternatively they had a really nice dish of grilled bacon and hot cheese with a little milk over it, which flowed in a nice tasty thick sauce, with plenty of bread. Beats all your pizzas!

Going in for the afternoon shift, John Wiffen, partial to the sausages served in batter ('though the sign on the counter, "Battered Sausages" could be taken two ways') always ate in the pit's main canteen and then went to the small one to buy cream buns. 'I used to eat them in front of the miners, my face smeared with the artificial toothpaste-like cream. It was fun to hear a scandalised collier explode: "He's had three bloody cream buns as well as his snapping."'

Before a shift, Arthur Gilbert always called into the canteen for a pint of tea, half asleep, as often as not in his pyjamas under his pit clothes; he had two alarm clocks (each in a separate biscuit tin) that invariably failed to wake him and it took passing miners banging on his door to get him up. 'They usually had to wake me in the canteen too, to catch the last cage down,' he says. He was back after his shift for 'the customary pint of tea with pork pie before bus time'. The German Victor Simons developed a liking for Horlicks ('It was new to me then') but the majority of Bevin Boys downed copious amounts of milk, trying to rid themselves of the dust; the national weekly milk ration was 2½ pints a head, but in pit canteens they were able to buy as much as they wanted. Most Bevin Boys felt ravenous and succumbed to the temptation of a slice of 'dip' (bread in hot bacon fat) when they could afford it.

At a number of collieries, miners given the choice of pithead baths or a canteen opted for the canteen, a choice that Reg Taylor at the Yorkshire Nutters thoroughly approved of. 'The pit manager's wife ran it and a grand job she made of it,' he says. 'The meals she put on were remarkable. Most of the miners ate there.' That appears not to have been the story all over the country. Most miners went in for tea or items like twist tobacco, but perhaps half refused to eat cooked meals in their canteen. They rationalised their refusal by saying that if supplies didn't go to the canteen, there'd be more available for the domestic ration – illogical in that if they'd eaten in the canteen, the food at home would have gone further. Throughout the war the miners' Federation called for extra rations for miners' homes rather than subsidised canteen meals.[12]

Some older miners refused to use the pithead baths, though this was on health grounds rather than point of principle. They believed that hot showers would weaken them and leave them open to chills. The attitude bemused Michael Edmonds in Wales:

The stigma of dirt of a thousand men could be washed away under a hot shower in an hour. The baths at Bedwas cost £42,000 prewar but some still used the tub at home before the fire. Imagine walking home with a wet shirt on one's back, limbs growing stiff in the chill of winter. They were frightened of skin diseases. A few believed that some sort of parallel existed between the removal of their dirt and the rape of Samson's locks at Delilah's hand.

*　　*　　*

If men at collieries with pithead baths hated 'getting into their muck' at the start of a shift, those at collieries without pithead baths hated even more travelling in it at the end of a shift to their homes, hostels or lodgings. Non-bath collieries had outside water taps for men to wash their hands and faces if they wanted, though nothing to dry them with; the Clifton colliery in Nottingham, where a number of Bevin Boys were among the 500 workforce, had a single tap.

Travelling in a pit bus or coach was one thing, public transport another. In some places miners traditionally stood, so as not to make the seats dirty; in most mining communities where pit dirt was a recognised fact of life, they could sit. It didn't mean they were comfortable. Roger Spencer, a railway clerk from Richmond in Yorkshire, who was at Tursdale colliery in County Durham, went 3 miles on a bus to his hostel at Ferryhill and 'when I got up, I'd leave a little pile of dust behind. Regular miners were getting on and off, so were other people who were probably miners' relatives. But I still felt ashamed.' So did Syd Walker (the pit bus first, then two public buses and finally a tram), on his way from Great Wyrley to his home in Huddersfield. 'The sense of shame I felt is something I can still feel,' he says. 'I travelled "alone" because no one wanted to sit by me.' Adds Reg Taylor, a 4-mile journey from pit to home in Bradford: 'Many civilians weren't very civil – they didn't like dirty miners sitting next to them, even though more likely than not they were wearing overcoats or raincoats over their working clothes. After a number of complaints to the bus company the colliery laid on a special bus.'

In many cases Bevin Boys walked to and from the pits, often 3 or 4 miles, which taken with an average of 2 or 3 miles underground to and from work stations made 30 or 40 miles a week. Missing a bus getting out of the pit late on afternoon shift and missing the laid-on transport or the last public bus meant more walking. On one occasion Doug Ayres missed the hostel bus from the Homer and Sutherland and

walked six miles into Stoke and stayed the night with the Salvation Army before doing the next six miles the following day. At three in the morning we were visited by the police looking for deserters, but I had my identity card. It was after that experience I got digs locally!

Working nights, when there wasn't any transport, was a problem. Rather than walk the 6 miles from his hostel when he was on nights at Blagdon near Newcastle, Ken Sadler preferred to catch the last tram at 10p.m., which got him there at 11, and hang about in the canteen for three hours before going on shift. Tursdale colliery was only 3 miles from Ferryhill and when on nights Roger Spencer walked it regularly.

> In 1944 there was still a blackout. One night in thick fog I came across a white apparition. I honestly thought it was a ghost. I absolutely froze, then found myself starting at a large carthorse which had escaped from a field. Another time, walking with another Bevin Boy, a Scottish lad, I was knocked down by an Italian prisoner of war on a bicycle under a railway bridge. He was returning to his billet on a nearby farm. I don't know who was more startled, him or me. I landed flat on my back. He landed in the hedge on the other side shouting 'I am hurted, I am hurted.' In fact he was all right, but his bike was done for – he had to wheel it away on the back wheel. We said we'd get some assistance, but he didn't want that. He was seeing a local girl and shouldn't have been out.

Living at home and needing to remove the pit dirt after a shift, Reg Taylor steadfastly ignored the government's exhortation for everyone to bathe in no more than 5in of water. Indeed, he not only had a full bath, he had two. 'That's what it took to soak away the pit dirt,' he says. 'It took coal to heat the hot-water boiler and I was down the pit providing coal. I had no conscience about it whatsoever.' Three years of soaking away the pit dirt, Geoff Darby reckons, 'ruined my parents' bath'.

It would be a mistake to assume that all Bevin Boys who were able to live at home had the luxury of being able to soak in a bathtub. Ken Tyers, a general labourer before conscription who every day cycled 10 miles in both directions between Wooley colliery and Tow Law, Bishop Auckland, lived in a house that had 'no electricity or plumbing. I had to wash my hands, face and feet as best I could in a small dish of cold water from out the back. At weekends I washed the other parts behind the pantry door.'

For most Bevin Boys in digs, washing meant a tin bath in front of the kitchen range, like generations of miners. Some, lodging with a miner's family, only got to use the water after the miner had finished; others, who lodged three or four together, had to share the tub. Not all miners bathed every day: David Reekie's landlady only got out the tub for her miner husband on Fridays. Reekie would have nothing to do with it. 'I was always a shy person,' he says. Daily he returned and washed less than properly 'at the sink in the corner of the living room'; on Fridays he was able to go to the home of one of the two colliery handymen who'd befriended him and use the bathroom.

For the rest of the week I wasn't clean. At times I dreaded going out. At chapel socials they used to have novelty dances, roll left leg up, that sort of thing, and I didn't take part; often my legs were as black as ink. At someone's house a woman said to me in a loud voice, 'You poor dear boy, among those horrible men.' I grew angry as I still do when people are superior on a class basis. I rolled up a trouser leg up and shouted, 'This is what those horrible men have to put up with. There's no facilities, no toilet, no baths, nothing.' I shocked everybody.

To their initial consternation, many Bevin Boys who used the tub in front of the fire found that their landlady scrubbed their back. Geoff Rosling returned from his first day down the pit 'black, sweaty and filthy' to discover

Mrs Walters filling a long tin bath with kettles of hot water from the cooker. She must have been working hard at this for some time. Hobson's choice meant I had to strip off and get into the bath. I knew it was top half first. This done I stood up to deal with the rest when Mrs Walters came in with another kettle of hot water. She didn't turn a hair, unlike myself. Later she told me how her miner Da had always washed in this way and she often scrubbed his back while Mam was cooking.

Alan Gregory, lodging in the house of a surface worker at another colliery, also had his back scrubbed by his landlady, 'in a motherly way'. So did David Roland who was

billeted with an old schoolteacher who had her nephew and his wife living with her. He worked at the pit on the surface, there was something wrong with him. His wife used to wash his back in front of the fire and she washed mine too. But unlike him I kept my underpants on and then went upstairs to wash downstairs, as it were.

Ken Da Costa didn't get his back scrubbed; his landlady filled his tub and left him to get on with it. 'But the local girls knew what time I'd have a bath and started to come in to have a look. It was dead embarrassing at first. After that – well, say no more.'

Five

Hurt or Worse

One thing that all Bevin Boys learnt in training, whatever their enthusiasm or lack of it: if you didn't do things right you could get hurt. More than a few suffered mishaps. 'In a dormitory of twelve Bevin Boys,' recorded John Platts-Mills, 'we had three with broken bones in the hand in a fortnight'.[1] Some, like Doug Ayres, struck on the shin by a tub, were injured badly enough to be sent home for several weeks to receive outpatient hospital treatment. However, at the end of January 1944, the newspapers were able to report that 'in the first month of training, no conscript miner has been seriously injured'. The government's relief was implicit: it was, after all, sending non-coalminers into the industry with the worst safety record in Britain.

In 1943, the year before the Bevin Boys went down the pits, 713 miners had been killed in underground accidents – a figure below the 1939–45 average, which was nearly 900; in the three years 1943–45, 2,353 men suffered serious injury. Roof falls accounted for just over half these totals; haulage accidents for something over 20 per cent. Explosions and fires were responsible for under ten per cent, but as individual incidents were the most destructive. It was sobering for Bevin Boys to serve in a pit where a major disaster had occurred. John Wiffen at Gresford in north Wales, where 264 men and boys died in 1934, 'near enough in time for mention of Gresford in the neighbourhood to cause averted or lengthened faces', the place where the bodies lay was sealed off and 'men passed where their fathers or other relatives lay every day – twice. I got used to a jerked thumb in the direction and such comments as "My two brothers are in there".'[2]

The general belief is that the safety record of mining worsened during the war as equipment disintegrated and regulations were skirted, sacrificed to the single objective of getting coal up the shaft to drive the war effort.[3] Many Bevin Boys became aware of contraventions of the Coal Mines Act. Eric Gulliver, who volunteered for the mines after being invalided out of the RAF, noted that at the Clifton colliery in Nottingham, refuge holes on haulage roads were less frequent than the stipulated 10yds – the estimated radius of a man's lamp – and, in breach of the regulations, were absent in places where the gradient was more than one in twenty; they were also often in a bad state of repair and obstructed by piles of wood or metal. No inflammable materials were supposed to be stored underground, but drums of oil were, and packets of gunpowder for shot firing left in a corner. Before call-up an investigator for Mass Observation,[4] Gulliver made it his business to talk to Bevin Boys elsewhere and found similar infringements were common. On occasion when an official's safety lamp indicated a dangerous level of gas[5] and work should have stopped and men withdrawn, a reading just over the threshold was often ignored – although fresh brattice would be rushed to deflect more air upwards to disperse the danger. There were also instances of shot firing being conducted in pits where inflammable gas had been detected on the face within the previous three months, in direct contravention of the rules. Even stone dusting, which reduced the risk of a coal dust explosion – a far greater risk than a gas explosion, being far more violent and widespread – was often skimped. 'For a normal mine roadway, not less than 50 per cent of dust should have been stone,' says Ian McInnes.

> Samples were regularly taken and if it wasn't a colliery could be prosecuted. Mine inspectors came at half an hour's notice. I reckon that if the inspector as he walked the mine with the manager had hurried up a bit he'd have caught two blokes scattering lime flour like bloody mad.

Another safety regulation that many collieries did not observe because they could not was the issue of a safety lamp to a given number of the ordinary workforce (usually to every tenth man) in addition to their own; wartime shortages reached a point where there weren't sufficient serviceable safeties. Most pits were still using 'lighthouses' and these caused problems, too. Cap lamps had begun to replace them in the mid-thirties but the war had put a stop to that, and the stock was old and unreliable.[6] 'You could almost lay odds they'd go out while you were using both hands in a tricky situation,' – Stan Payne's observation. Some collieries were even forced to issue obsolete oil lamps, which were brighter than lighthouses but prone to going out if they got a sudden knock; not being relightable without being opened, potentially dangerous underground, they had to be sent to the surface to be relit, wasting considerable time. (Some pits kept an ingenious battery operated gadget for relighting lamps but, from what Bevin Boys remember, it never seemed to be working.)

A lamp suddenly extinguished could be a frightening experience. 'It felt as though the blackness came right up to press against the eyes,' says Desmond Edwards. When John Wiffen knocked over his lamp and it blinked and failed, he found out 'what it was like to be blind'. Les Thomas, on his way alone to pit bottom at Yorkshire Main, suffered a lamp failure and sat in a refuge hole for two hours before an electrician happened by and 'I came out and gave him a fright – he nearly jumped out of his skin.' Ken da Costa gave the lamp man at the Arkwright pit near Chesterfield more than a fright after being the victim of a lamp failure for the third or fourth time.

> When I got up I was raving. I got the odious little creep by the throat and told him: 'Do that to me again and I'll kill you.' It was completely out of character. But I was scared and angry. He just didn't look after the equipment, he hadn't charged the lamp. And the bloody thing burnt a big hole in my overalls. I didn't get into trouble: some of the miners had had a go at him over the same thing.

At least most of Bevin Boys worked in pits that had lamps: a few, like Brian Evans, worked in pits that only had candle power and 'you couldn't see at the best of times'. Evans adds:

> You went down with eight or nine fat tallow candles and you put one in a half tin that you could hook over the tub in front of you so you could see your feet, or at the front to see where you were going. Bloody things went out all the time and you had to feel your way along the rail like a monkey, pushing your way through the overlapping brattrice sheets on the other side of which, invariably, some bright bugger had left a tram [tub] and bong!, you knocked yourself silly.

It didn't help that like all the regular miners in the Beech Tree pit near Stourbridge, Evans chose to wear a beret instead of a safety helmet.

Not until the 1950s was it compulsory for miners to wear helmets underground and many didn't, preferring a beret or, in more cases, a flat cap. The flat cap particularly was a long tradition and miners were essentially conservative, reluctant to give up the old ways. But there was more to it than that: some never articulated feeling was that helmets were unmanly; miners saw themselves in combat with the coal and took it on with an air of defiance, on equal terms as they seemed to see it. A few Bevin Boys who adopted caps or berets were perhaps moved by the same spirit. Norman Brickell did so simply because he hated wearing his helmet. 'Mine remained in my locker – nasty, uncomfortable thing made my head sweat in the heat. Even the colliers on the coalface never wore them.'

The overwhelming majority of Bevin Boys regarded all those who dispensed with their protective headgear with disbelief. Most removed theirs at snap – but not before looking carefully at the roof, in case a fall seemed imminent. Many took the precaution of sitting by the side of a tub; the very careful opted to sit in a refuge hole. 'I thought miners who didn't wear helmets were cocksure and daft,' says Brian Folkes.

They bashed their heads like anyone else. Even the main roads weren't always full height: considerable distances had to be walked bent double. It was all too easy to walk into a steel roof joist. If you were wearing a hard hat it merely felt as if you'd broken your neck – the back of the helmet smacked into it.

'Strange as it may seem, I was the only man who wore a safety helmet – but then I was the only Bevin Boy in the pit,' says David Reekie.

It never left my head and certainly saved me from injury on many occasions. The protection wasn't so much against falling rock but the far more threatening hazard of hanging pulleys and roof bars. And the old hard hat was jolly useful when the only way forward with a loaded tub was to lie in an almost prone position and push with your head.

Hazards abounded in the pit from the cage to the coalface, exacerbated by bad light in most areas and choking dust in some. Stone on the side of roadways and the battered edges of tubs were often razor sharp. The strands of the endless moving steel haulage ropes frayed into nasty spikes and splinters that could be avoided by those clamping tubs to them if they wore gloves, which some Bevin Boys did but were advised not to as 'feel' was critical: it was all too easy clamping on the run as tubs were pushed away to clamp a gloved finger.

The tubs themselves were what Bevin Boys working on haulage had to be most wary of: it was all too easy to get between a stationary one and another that stopped only by crashing into it. But 'runaways' were the constant real danger, caused in many ways. By a haulage man missing his clamp ahead of an incline or decline. By a link unshackling under the strain of drawing a set of laden tubs up an incline. By an engine failing. Or a rope snapping. When such incidents happened, gravity was the enemy.

The cry 'Runaway' or 'Runner' or 'One away' ('Awain', in County Durham) was to be feared: a tub or a set of them, empty or laden, was coming rattling out of the darkness and men dived for the refuge holes. Invariably runaways ended in a pile-up and, as John Wiffen witnessed, 'Girders sprang out, chains snapped, rails were bent like thin wire, and full tubs wedged against the roof. There was danger all round the pit as the rope lashed.' In a newspaper interview years afterwards, the late artist Tom McGuinness, a Bevin Boy at Fishburn colliery in County Durham, recalled:

The first day I had a supervisor with me and about two hours into the shift I heard a noise coming down from inbye. The man with me said the tubs are awain and we must take cover in a refuge hole nearby. We just made it into the hole when the tubs went speeding by. I attempted to go out when my supervisor said stay put as the tubs would come back and so they did, running backwards and forwards until they finally stopped. The man at the top had missed with his dregs [lockers]. This was apparently a regular event and on one occasion 26 tubs got awain and this time I could not make it to the refuge hole, so I just laid in the side and hoped. Fortunately the set got off the way just a few yards from where I was. There was coal all over the place, squashed tubs, my light gone and the signal wires cut. One miner on his way out said to me, 'Son, you have a worse job than a rear gunner.'

Roof falls were much less frequent than accidents along the haulage roadways but much more likely to cause death or serious injury. And they were common enough. All roadways underwent what miners termed a closing or a squeeze: the slow, relentless pressure in from the sides, up from the floor, down from the roof, 'as though,' as Desmond Edwards puts it, 'nature resented the existence of an artificial cavity defiling its innards and was determined to fill it'. Roof falls, however, most of them on the coalface, were sudden and could be calamitous.

There were more coalface cave-ins during the war than before it, because good timber was scarce and collieries took chances. Before the war, pit props were of Canadian or Scandinavian timber, 'clean and straight as miniature telegraph poles' in Geoff Rosling's words. When stocks became depleted, however, British wood, usually deciduous hardwood, was substituted and miners considered it downright dangerous; the imported timber creaked and groaned as it gradually accepted the weight of the lowering roof – and a change in the sound gave a few seconds warning before it cracked. The pit props talked, miners were fond of saying. British timber, 'unseasoned, crooked and gnarled, and still with the leaves on', in Syd Walker's memory, simply snapped, often on the day it was erected. Some collieries tried small girders, which were even more unpopular. Not only could they not be cut to size, they were as often as not forced out of position – and they broke up the roof, too, creating rather than preventing a potentially dangerous situation.

Until the war pit props had generally been left in the gob to be destroyed as the stone packs were crushed by the settling of the earth. Now colliers tugged some of them out to reuse as they worked and, as Alan Gregory observes, 'occasionally they carried things too far'. In Geoff Rosling's pit 'a "save and rescue" system was in force on the night shift, which was where Tom Jones "the Buller man" came in.'

The Buller was a ratchet-actuated method of extracting precious props by anchoring one end of a chain to a firmly fixed post, and the other to the prop to be saved, then moving the long handle back and forward to tension and pulling the prop out. There was considerable skill in knowing which props to tie up to and how many could be salvaged without causing a dangerous collapse. Sometimes Tom for all his skill got it wrong. When he yelled 'Get out', you didn't hang about.

All Bevin Boys who worked as colliers learnt, like Desmond Edwards, 'to test the top for soundness, tapping gently with the mandril [pick] and at the same time lightly applying the other hand to feel for any bagginess, any fragment or slab that might come loose'. Not that doing so was any guarantee, as Jack Garland found out, almost to his cost:

Our seam was a little over six feet but it had a bad roof, roughly three feet of razor-edged shale, so to support it we left about ten inches of coal at the top. On one occasion the roof, according to my butty, 'rang like a bell' and he decided we could take the coal underneath – easy coal. We brought the coal down but while shovelling it away I felt a dribble on my back. Then there was a wrenching sound, like ripping cloth. I sprang to the side of the road and the lot landed in front of me. We never tried that again.

Frank Pratt's service in the pit lasted only two months: his MP raised his case in the House and he was released to look after his invalid mother[7] – but not before his butty saved his life on the coalface.

The wheel of an empty dram [tub] being brought in struck a roof support. There was a silence. Then a thin run of earth started coming out of the roof like sand in an hourglass. I just stood looking at it, it meant nothing to me. Trefor said run. I ran. The roof collapsed. If I'd been standing where I was I'd have been killed, no question.

Ronald Griffin, another Bevin Boy collier, was saved not by any physical warning but a sixth sense.

I was heaving coal on to the conveyor when something told me to move. I felt there was someone behind me. Lunging forward I cannoned into my butty just as there was a tremendous crash followed by clouds of grey dust and a pile of stone rubble. If I hadn't listened I'd have been under it.

I used to attend spiritualist meetings before national service with an aunt and the medium, a marvellous woman, asked if I was afraid of going into the pit and told me I had nothing to be afraid of, I'd be looked after. Mind, she also told me she could see a globe of the world over my head, 'and travel you will', she said. But she wasn't wrong about that, travel I did – down the Aberbaden pit.

It was something that Syd Walker remembered from training that probably saved his life and that of another Bevin Boy. They were working on a haulage road when Walker

> heard a sound like pins dropping, which I'd been told was a warning that the roof was about to fall. I turned round to see a steel girder moving down. I grabbed the other boy, pushed him under it through the mud and followed him. The roof fell and the girder hit my heels, but I was out and we ran and ran until we reached the main road, absolutely terrified.

Almost every Bevin Boy remembers at least one occasion when he had a narrow squeak. Derrick Warren heard the roof go when he was on a night shift at Easington colliery in County Durham and

> came out scrabbling on my hand and knees, the lot coming down behind me. The man who every night unscrewed the teeth from the coalcutter to sharpen them had left his toolbag and was intent on going back for it. When they couldn't restrain him, one of the miners knocked him out with a punch on the jaw.

Doug Ayres at snap had 'just gone over to sit with my mates when the roof where I'd been standing by my twisting plate came down – a large lump of rock, about five feet long and two feet thick. We were quiet for almost a minute.' Geoffrey Mockford was walking to pit bottom at loose it and had gone a mile,

> when there was a great rumble, a crash and within a few seconds the roadway was full of thick dust. I took to my heels but after a few yards things quietened down. Five yards behind where I'd been was a hole in the roof and a pile of rock six to eight feet high covering the whole width of the road. I wondered at my luck. What if I'd walked a bit slower, or started back a bit later?

Coming to a roof fall several yards high along a roadway, Desmond Edwards climbed over it. Some distance on he passed an official going the other way, who at the end of the shift asked him what the fall had been like when he'd gone by – the official found it totally blocking the road. '"It had all come down like a cathedral", he said, and I thought how chancy life is.' Chance saved Alan Gregory. 'One day I was working as a motor driver on the gate belt next to the coalface. The next I was moved – and the gate fell in on my replacement breaking one or both of his legs.'

For the majority of Bevin Boys the great fear wasn't about being injured in a fall but being entombed by it. That happened to Bert McBain-Lee, a few weeks before his service ended. He'd been given an opportunity ('which the pit manager said most lads would give their right arm for') to work on the surveying team and study part-time for a degree in mine surveying, and he was with the head surveyor in a 3ft return road using a theodolite and sighting lamps to check its alignment 'when there was a mighty crack and a rumble – a roof fall ten yards behind me'. McBain-Lee tried going ahead but 20yds on another fall blocked his escape that way. He panicked.

I was in a tomb of approximately 30 cubic yards and had no idea whether the surveyor was trapped or not. I thought I might die. I wondered who'd tell my mother. The first thing I did was go at the rock with bare hands and nails, but I hadn't a cat in hell's chance of digging myself out. And I remembered a lecture about conserving oxygen. So I lay down and awaited delivery. I was a stable-minded person. It was eight or nine hours before I heard the sound of shovels and another hour or two to get me out. I was given a shot of brandy and carried out of the pit on a stretcher to the medical room, but no damage – except to the manager's ideas about my future. Thanks, but no thanks.

To many Bevin Boys, irrationally perhaps but understandably, the cage was the most hazardous thing about the pit or certainly the symbol of all the hazards below. Up to the early years of the century death and serious injury in cage accidents were common; by the 1940s, they were rare. Winding cables were capable of holding 100 tons and were checked daily by a fitter passing them slowly through a galvanometer; and they were replaced every three years, even during the war. Cages were also fitted with safety devices. If a cage was going too fast as it approached the surface and was in danger of going out over the winding wheel, a braking mechanism cut in to halt it; if the cable broke and the cage fell, clamps operated to grip it.

To be in the first situation was a heart-stopping experience, as George Ralston found out at Newtowngrange. As the ascending cage neared the surface it suddenly seemed to drop. Ralston didn't know whether the cable had snapped or the winding engine had failed. All he thought was: 'This is it' – and wondered at the silence of all the men around him. Then the cage steadied and rose, and Ralston realised what had happened: 'Men were in a hurry to get to a local football match and the windingman was getting a move on – too much of a move on.'

To be in the second situation almost didn't bear thinking about, but Syd Walker thought about it all the time and, what was worse, what would happen if the cable broke when a cage was descending.

We'd been told that going up was 'safe' because of the clamps. It meant a sudden stop, which would throw you up against the roof, but at least you'd be alive, even though injured. Going down you wouldn't be so lucky. The only thing you could do was bend your knees at the moment of impact, though how we were expected to know when we reached the bottom no one could tell us.

He trusted nothing about the cages. 'A pit 3,000 feet deep meant 6,000 feet of cable on two cages. They checked them, but I was never convinced it was exactly a close inspection or that the instrument would know whether strands in the cable core were busted.'

Harold Gibson was no more concerned about going in the cage than most other Bevin Boys but had an aversion to it stopping halfway down the shaft – which happened to him on his first drop in training; that was when he discovered that if a descending cage stopped before settling on the solidity of the pit bottom chocks, it bounced up and down like elastic. At Huncoat, therefore, he avoided the maintenance man with the hinged plank.

On one occasion I was going down when a chap got in with what seemed like two 12-inch planks under his arm. To my horror the cage stopped when we were partway down and juddered up and down. That was bad enough. But then he lifted the concertina side of the cage so there was no protection at all, opened the planks, which I then realised were hinged, and calmly walked across to a heading to check the pumps. I was clinging to the bars with white knuckles. Whenever I saw him and his hinged plank I let the cage go and waited for the next.

Just as the push for productivity saw safety regulations flouted down the pits, so were some relating to the cages; their gates were often not working and even when they were men didn't always bother using them. Peter Allen at Ryhope was aghast. Even when the gates were operable 'they were insufficient and you could have slid underneath. Grease from the tub axles made the floor slippery. You had to watch your step. The cage was a dangerous vehicle.'

Cages could certainly be dangerous to be near at pit bottom when coal was being wound. From time to time a tub failed to connect with the docking mechanism inside a cage and bounced down the shaft or, if partially sticking out if the winding signal was given too quickly, struck the pit bottom roof, incidents that not only had men scattering in all directions but caused damage that took hours to repair. An accident in the main shaft that wasn't cleared by loose it meant the shift had to get out of the pit via the cage in the return shaft. Arthur Gilbert remembers that happening and 'hundreds of cursing miners trudging up the one-in-three gradient to the return in a tail wind and dust storm'.

He had reason to be nervous about travelling in the return shaft cage, which he often did when he fetched up in that part of the pit after inspecting the back roadways in his job as pit safety officer. 'There wasn't an attendant at the shaft bottom so I had to communicate with the winding house myself to say I wanted to return to the surface,' he explains.

> The method of doing that was pure Heath Robinson. By the side of the cage was a yard-long steel bar with a handle. This was hinged to the wall with an attached cable up the 550 yards of the shaft to the winding house. I pulled this handle three times, which rang a bell to tell the winder I wanted to come up. The operation required a great effort and it was necessary to swing off my feet and use my whole body weight to achieve each pull. I'd then rush into the cage and close the gate, hoping against hope that the winder had received my summons. If there was no movement after a bit I'd leave the cage with my heart in my mouth and repeat the operation. Talk about dicing with death.

Gilbert is fortunate to be here to talk about it: a moment of carelessness on another occasion almost cost him his life and he acknowledges he should have known better, just as earlier in his service he should have known better than to operate the levers of his haulage engine with his boots.

One of his regular tasks was to take a sample of air in a return roadway notorious for blackdamp, which had been 'stopped' at either end by a brick wall following a fire. His usual procedure was to first check for gas with his safety lamp but 'familiarity had lessened my awareness' and one day he didn't bother. A little distance in he fell unconscious. Fortunately for him another safety precaution was being observed: he was only allowed to carry out his inspection when accompanied by the pit bottom fireman who was standing watching him as he went in. As soon as Gilbert went down, the fireman raced in and carried him out – and when Gilbert came to with no recollection of what had happened, swore him to secrecy. 'The fireman's name was Jim Gould and he said he'd be in trouble because he'd broken another regulation – he was meant to fetch breathing equipment. But if he'd gone for that I'd have been gone. I never met him away from the mine but I've never forgotten him.'

The repetitive nature of much of the work, and the heat, induced a somnambulistic state that induced carelessness. David Day, shovelling 'an interminable river of slack', overcame boredom 'by entering into a semi-trance whereby my body automatically performed the function of shovelling, leaving my imagination free to rove.'[8] Once, the deputy fireman ordered men back from the coalface for a shot firing that failed to explode. When he went to the face to check his wiring, he found Day there, still shovelling. Peter Allen, so wary of the gates of the cage, on several occasions was less observant about the haulage rope, a far more lethal threat, and almost paid for it.

The rope would lie quietly on the floor when it wasn't hauling and I'd forget about it, which wasn't very clever. When it came under full tension at bends where it ran round a series of steel drums you didn't want to be on the wrong side of it, and you didn't want to be across it at drifts where it hit the underside of the roof like a whiplash. I had one or two very near misses through lack of concentration.

Falling asleep was the big risk for those whose jobs were intermittent and could have serious consequences, especially when it involved machinery. Driving a haulage engine on a quiet night shift, George Poston sat down on the bench in the engine room and nodded off. He awoke 'to the loud roaring noise of a full train of tubs careering down the incline past the door'. He leapt to the controls and stamped on the foot brake, but the engine seized, the revolving drum stopped abruptly, and

the impetus of about 30 tons of coal running away down the one in nine incline almost wrenched the engine from its mountings. Something had to give, the rope snapped and came snaking back into the engine room to coil itself around the central air pipe just above my head. Two feet lower and I would have been garrotted.

The tubs continued another couple of hundred yards to the bottom of the incline where they piled up in a heap of coal and broken timber and brought down part of the roof. It took Poston and the rest of the men twelve hours to dig themselves out and he

was expecting to be put under arrest when the mine manager finally broke through. But before he could speak, the deputy, who'd already given me the biggest rollicking I have ever had to endure, tore into him about the rotten standards of maintenance and that the accident was all the fault of management for providing old worn-out ropes and for not maintaining the haulage engines in a proper manner.

The engine was overhauled and fitted with a new rope, and Poston kept his job, 'but the deputy made it clear that if there was a next time I was on my own'.

Carelessness could endanger other workers, not just the individually careless; so could practical jokes, which sometimes weren't harmless as intended. Before his entombment, Bert McBain-Lee was twice a victim. It was critically important in mechanised pits that the system of bell signals – one to stop, two to start, three for an emergency stop, all made by touching together two wires that ran along the roadway – was adhered to, and that only the man who'd stopped a rope started it again. In the first incident, McBain-Lee was uncoupling tubs when someone else started the rope without being signalled and a link snatched the top off his ring finger. 'The doctor at the hospital said that he thought the bone above the top knuckle was still complete, but there would be no new growth of flesh and no nail,' he says. 'Luckily, six weeks later when he took off the dressing there was new flesh and sign of a new nail.' In the second incident McBain-Lee was sent on his own to bring off a big run of full tubs, sending six at a time a hundred yards apart, along the main roadway. In the space between the 'full' and 'empty' tracks,

chains were coiled with the hook on the top, lying in about six inches of dust with only the hook and maybe the top coil visible. Unfortunately for me some silly so-and-so had deliberately coiled a chain, which was still fastened to the hook of an empty tub and led under the empties, so that when it stretched its full length and tautened it pulled over a full tub on to its side. A half ton of coal pinned me down between the tracks and squashed my right thigh. As there were two sets of brattice sheets suspended in the roadway no one heard me or saw my lamp movements, and because the doggie [haulage foreman] thought I was skiving no one came to look for me for about two hours.

That one kept him off work three weeks.

John Wiffen worked with a young regular miner sending empty tubs from one station to the next, who thought it funny to lash several sets as close as he could or spread shot firer's clay on the front and back lashings, and couldn't see he was causing potential danger to the receiver. Wiffen:

The foreman had a way of dealing with Maldwyn. He would invite us to sit down on a rough bench made of sleepers and pit-prop ends. He would sit between us and carefully put down his roadman's pick. Preliminary conversation would include such harmless topics as the weather and the local football results. But I got to know what was coming next: he would reach over Maldwyn's leg and grab a handful of loose and tender flesh on the inside of one leg, just above the knee, and squeeze. At the same time he would ask, still in the most civilised tones, 'What made you send all those tubs in together like that?' Maldwyn was unable to reply as his face was contorted with pain. Eventually phrases like 'I'll never do it again' were wrung from him. But he didn't learn.

Pits issued cards of safety instruction making clear what practices were banned, but that didn't stop miners and Bevin Boys alike from ignoring some of them. Almost everyone now and again hitched a ride on a conveyor or rode between full tubs up drifts (very dangerous, even though the engineman knew and took extra care); in pits with undertub haulage men frequently stood on the clips between tubs and went to pit bottom that way. Occasionally men were forced into doing something illegal by circumstances. Geoffrey Mockford was always at pains to obey the rules but was stranded along the roadway when his lamp failed. 'I could hear full tubs coming along, so I waited well up to the wall,' he says. 'They passed me – I put out my hand and felt them as they passed. Sound helped to tell me which was the last one in the line and grabbing hold of the back edge I stood on the clip. What else could I have done?' Harold Gibson makes a similar point:

> At Huncoat endless chains were used, a link dropping into a fork on the tub. This was very easy and effective until there was a dip in the road, which caused the link to come out of the fork. On an incline this meant a run-back – and of course you got out of the way of that. Then someone had to re-rail the tubs and take a train of 15 or 20 up the incline by kneeling on the chain of the last tub to keep it tight. A safety officer today would have a fit.

There were no circumstances in which a putter could be excused for riding a pony and if he was caught he was heavily fined, but many did, including Norman Brickell. Not all ponies would let men ride them, but 'Dandy didn't mind,' Brickell says,

> though despite my urgings to go faster he only went at his own pace during the shift. But at the end of it, along the last mile and a half to the stables down a little-used back airway, he was a Derby winner. I lay flat on his back with my head along his neck and we'd streak home. It was dangerous, there were some very low girders in places, and Dandy only had one eye. But I was young and stupid – and it was better than walking.

*   *   *

Collieries generally looked after their ponies which was, of course, in their interests. Some, however, overworked them. The finger-and-thumb rule was that a pony should only work one shift in twenty-four hours and never more than three in forty-eight. But as the Pit Pony Protection Society complained, there was no legal limit and, as many Bevin Boys, like Jim Ribbans at Werfa Dare in south Wales, observed, back-to-back shifts were common.[9] The harsh reality in Pelton Fell, Durham, Phil Robinson says, was that 'the good ones were worked to death, the bad ones got away with it.'

Before the war the RSPCA had pressed for legislation for pit ponies to be retired at fifteen but many were as much as twice as old. According to the Mass Observation Bevin Boy Eric Gulliver's research, a conscript miner in a Welsh pit was so incensed that a pony named Avery, who was reckoned to be thirty-two, was worked three shifts without a break that he anonymously contacted the RSPCA, which negotiated the pony's freedom and retirement to a farm.

What was worse that overworking healthy animals was working ill or lame ones at all. According to Gulliver, a Bevin Boy at Bestwood near Nottingham told him the pit had many ponies in poor condition and when a government inspector came they were taken to a disused area where he wouldn't see them. The Pit Pony Protection Society prosecuted collieries before and during the war and sometimes individual miners – one for hitting a pony in the mouth, another for pulling out a pony's tongue, another for killing a pony. Such incidents were rare. But lesser acts of cruelty were frequent. Ponies were well cared for in the stables, but how they were treated once they were taken out depended on their drivers. Some regular miners were callous and Bevin Boys saw them brutally hit animals when they were unable to move a heavy load. Gulliver himself at Clifton saw a pony that failed to draw a tub forward 'given two vicious punches in the kidneys, which caused it to spew straight away'. Jim Ribbans 'often saw ponies struck with sprags [lockers]'. So did Derek Thompson. 'The colliers were on piecework,' he says. 'Their attitude was, we've got to earn our livelihoods – the animals were just their tools. I saw one or two burnt with carbide lamps to hurry them up. I complained but it didn't make any difference.' Norman Brickell takes a more detached view:

There was cruelty, not generally, but there was, I have to say that. I was on nights once working with a group of three men driving a new roadway. This was hard work and the pony wasn't adequate for the job. The men's wages depended on yardage that they drove and the poor old pony was badly beaten. Again, I was young, I don't think I felt particularly sorry. There was a job to be done.

Coalmining was a hard industry that hardened men, but few regular miners tolerated out-and-out cruelty. When he was in training at Pontefract, Geoff Darby saw a miner 'who was ill-treating a pony, dead nasty, kicking it to the stable, attacked by other miners who knocked him down – the only fight I ever saw down the pits'. If Darby had been at Chilton in County Durham he'd have seen plenty more: Dennis Fisher got into 'more fights than I can tell you' over the mistreatment of ponies.

Fisher had a series of confrontations in his first job in the stables with

the yackers among the regular miners – we'd call them yobos today – who used to come in, grab little Titch the runt of the litter by the tail, and lift his back end up, showing off, feats of strength like. Okay, pal, do that again and you'll answer to me. I was always in trouble but okay – the management cared about the ponies, I'll give them that.

A more serious incident followed when Fisher moved to pony putting, and it involved the death of one of his two favourites, Paul.

> A putter called Tubby took him out one day and the roof started closing on the landing where the full and empty tubs were standing. Tubby dived to retrieve his prized putter's cotrell – his had a ring on the end, which most didn't.[10] If he'd had the sense to lift the limber from out of the tub's yokehole at the same time, which would have only taken a split second, Paul would have been out and off down the roadway faster than Tubby. But he didn't, the lot came down, and Paul's bones are down there still.
>
> At the end of the shift going home – we travelled home dirty in them days – Tubby was up in front of me. I held him responsible, he'd only thought of his own neck. I caught up with him and knocked seven sorts of shite out of him.

Even when he moved on to the coalface, Fisher stood up for the ponies. When one called Jack that he himself had putted (he still has one of Jack's shoes, copper-plated) was being abused by the man now putting him, Fisher intervened.

> Every time this fellow brought Jack to the face and the pony wasn't turning round fast enough for his liking, he hit him on the head with a dreg. It was hard for the pony, the place was only six feet wide and with the limbers Jack was six feet long. This was going right through me and I kept telling him not to do it. The other collier I was working with as it happened owned a pony and lived in the same village as him and I said he should do something about it, but he didn't want to get involved. After I'd given this fellow many warnings, I grabbed him around the neck and with the flat of my hand I flattened him. 'Now you're getting what the pony's been getting,' I said. I picked him up three more times and flattened him again, and then my marra got between us.

Hardly surprisingly, ill-treated ponies were nervous and difficult, prone to rear or shy; a pony that took fright not only caused derailments but could bring down timbers, with the danger of a roof fall, or hurt someone in his path. Whether a pony called Twinkler at Chilton had been ill-treated or was 'a bit touched' Fisher didn't know, but

> he was one of the bad-tempered ones to look out for. When I worked in the stables and he was in his stall he never failed in trying to bite lumps out of my arse or stand on my foot. When he was working, every now and again he'd say to himself 'Enough is enough' and take off from his handler, charging between the sets and turning them over. There was no stopping

him until he reached the stables. For devilment the putter who used Twinkler would let him loose at the end of a shift and down the roadway he'd come, teeth showing, screaming, eyes blazing. The men at shaft bottom waiting for the cage scattered in refuge holes and if they couldn't they'd jump on top of the full tubs. The last time I saw Twinkler he was aged and following behind his handler like an old sheep, all the fire inside him extinguished. But his name lives on with those he put the fear of God into – and there's still a few of us about.

Whatever his pony's temperament, a Bevin Boy who putted him was grateful he was there if and when his lamp failed. Then he'd hang on to his pony's tail, head down against the pony's rump in case of overhead obstacles ('I was glad it was Charlie and not Jerry when it happened to me,' says Syd Walker, 'he didn't suffer from wind'), and the pony would lead him out. There were times when Bevin Boys had even more reason to be grateful – when their pony gave warning of an imminent roof fall. If a pony started to snort and stamp, or trembled, or refused to move, he was to be heeded. 'Listen to what the pony's telling you,' was the advice miners gave Bevin Boys, but it wasn't always easy to know if a pony had sensed danger or was just playing up. When George Poston's animal stopped in a very narrow and low part of a roadway through which they were taking a tub of spoil and refused to budge, he thought it was plain cussedness.

I tried urging him forward by pushing the tub into his rear and shying pieces of rock at him, but he wouldn't move. After five minutes of this there was a loud creaking noise and about a ton of rock fell out of the roof on to the track about 20 feet ahead of us. If the pony ever stopped for no apparent reason after that I was content to sit down and let him decide when it was safe to move on.

A pony even more certainly saved Dennis Fisher's life – and in the same incident he saved the pony's. It happened when Fisher was putting, on the first day they worked together, and he regrets that he can't remember the pony's name.

Me and my marra Ray had filled the tub, but the pony was reluctant to move, quivering, trying to catch his breath like. I said to Ray, 'The place is closing.' 'Don't be bludy daft,' he said, but the pony's right, the place is closing. Ray bolted. The pony couldn't get out. He was right against the coalface and I remembered from training that the safest place in a fall is right against the coal. I took all his harness off and his headpiece and stood his limbers up on the full tub like props. Then I crawled out on me stomach before the lot was down. I found Ray against the girders in a state of shock – he'd been a soldier at Dunkirk.

The manager decided to make a new roadway to get the pony out if we could – my dad was one of those set to dig it. It took three weeks. I crawled in and the pony was alive – he'd stayed alive by chewing the bark off the pit props. The head keeper crawled in to feed and water him until they got him out. There was a place on the surface, a whitewashed building, the pony hospital like, where ponies that got lame could recover. It was quite a while before that pony came back.

Miners liked to think that ponies acted to protect their handlers as well as themselves, but there was none of that with Jerry, Syd Walker's pony of all bad habits. Once, when Walker was in double jeopardy on the roadway – not just a runaway on the full track but another on the empty, Jerry beat him to the refuge hole. 'Obviously he was too big to go into the space and his rear stuck out over the rails,' Walker remembers.

It was a deadly serious situation – I had two lots of tubs coming towards me and nowhere to go. Fortunately for both of us there was a 'monkey'[11] between the rails on the full run, which derailed the tubs. We were showered with coal and the roof fell in further up the road, but we survived.

Many ponies didn't survive to reach a retirement field. Like Paul, Dennis Fisher's other favourite, Peter, didn't. Most ponies that died in the pits were crushed by tubs, Peter died

because they were working him in places too low for him. He hated having his back touched, so he used to bend his legs to pull the tubs. It broke his wind. They used the humane killer on him, like they did with the others that got crushed or broke their legs. They chucked them in an empty tub, at least one a week at Chilton. Pitiful.

If Fisher had his way he'd 'have a statue put in Westminster to all them gallant little animals'.

They worked up steep highs and down steep dips that really tasked their strength. They worked in the same terrible conditions as the men but more, aye. They got coal dust on their lungs and joint problems, just like the men. Being small they were close to the ground where the oxygen content was low and sometimes had blackdamp gas. They damaged their eyes and their hearing, breathed in shot smoke fumes and got thumping headaches. Without them, in Durham anyway, there wouldn't have been coal production.

*   *   *

Hundreds of men in the pits suffered minor injuries every day. The majority didn't keep them off work, but the number that was bad enough to keep them off for three days or more was significant. In the Durham coalfield alone in the year before the war it was 131,776; in 1944, the first Bevin Boy year, it was 173,716.

Bevin Boys endured their share of knocks and cuts, bumps and bruises. Fingers and thumbs were most at risk on haulage (Ernie Jefferies remembers one Bevin Boy in Glyncorrwg, a jazz musician who played with Stephane Grappelli, constantly moaning 'This is ruining my hands'). And backs. Scraping the line of the spine against a low roof could be excruciating. At one time or another almost every Bevin Boy was knocked flat in the dust crying out in pain. The lesson was quickly clear: crouch low and if you had to use effort to heave and shove a tub, don't arch and straighten the legs; but 'it took me a surprisingly long time to learn,' says Reg Taylor 'You'd think the first time you took the skin off the nobbles of your spine that would be a good enough lesson, but no, in the heat and frustration when trying to push reluctant tubs uphill, I did it time and time again.' Adds David Reekie: 'I was rarely free of a healing scab that I knocked off before it was ready to depart. Haulage was an ideal job for a tall masochist with a bony back.' In mechanised pits like Chislet, haulage hands learnt to cut pieces of discarded conveyor belting and cover the base of the spine and the top of the buttocks with it. 'An "arse-pad",' says Ian McInnes, 'was a valuable cushion.'

Many Bevin Boys left the pits with what miners called 'pitman's buttons': blue scars on the bumps of their spine. Many left with the same mementoes on hands and arms or foreheads, though in most cases these disappeared with the years. 'They were tattoos, in effect, the miner's trademark,' says Doug Ayres. 'Gashes and cuts filled with coal dust, which formed a layer over which tissue would form if you didn't apply a toothbrush in the baths to the affected raw flesh.'

Often, the only difference between a minor injury and a worse one was luck: if luck was against you, a bruised or split finger, for example, could instead be what miners called 'a burst'. Struggling with a 12ft prop through a 4ft passage, Geoff Roslyn trapped a finger, which 'literally burst out the top like a squeezed tube of toothpaste, only fuller and rounder and beautifully pink against the rest of my coal-black hand. The pain was beyond pain.'

The most serious risk on haulage – getting trapped between tubs, or between a tub and a pit prop – resulted in many leg injuries. Late for work one morning and anxious to get a string of empties to the coalface, George Ralston missed the eye of a tub with his clip and was pulled to his knees between the rails. 'By this time the hutches [tubs] were gathering speed – I could do nothing to stop them. I could feel the pain in my knees and legs against the rough rails and the ends of the steel sleepers.' He managed to pull free but his legs were a mess, he couldn't stand, and he was put to bed in his hostel's infirmary. He'd been dancing the night before. It was weeks before he went dancing again. Dave Moody fared less well: he slipped on a rail and had his leg badly fractured between a tub and a prop. 'I wasn't in that much pain,' he says,

> but the leg was pretty bad: they had to stretch it and they fitted a plate with three screws. I was in the Royal Victoria infirmary in Newcastle, then for three months in South Moor miners' hospital. I was sent home on crutches and continued outpatient treatment at Romsey – heat treatment, which was a bit of a joke, a big box thing of ordinary light bulbs which they lowered over my leg. Didn't do a damn thing.

Ken Tyers and Roy Doorbar fared worse than a leg injury. Getting trapped between two sets of tubs cost Doorbar a broken sternum, a bent spine and a chipped shoulder blade; getting crushed against a standing set of tubs by a full set that derailed cost Tyers a broken pelvis, a ruptured urethra and other bladder injuries.

Doorbar's accident happened because someone else didn't obey the safety signals.

> I was at the top of the dip linking tubs to the cable in threes before hitting the bell, but they wanted to finish winding coal and weren't giving me enough time to get the coupling links sorted. At least I got my head out of the way. We didn't have a doctor at the pit. We had a nurse. She got the ambulance but she didn't send me to hospital, she sent me home. If we'd driven up to the door my mother would have had a fit, so I got the driver to stop a hundred yard from the house and walked there in agony.

When his mother saw the state of him she had him taken to the hospital. He was in and out of it for two years, though he went back to the pit after a few months and worked in the repair shop mending conveyor belts and then in the workshop cutting pit props to size. 'They called it light work,' he says. 'Believe me, there's no light work in a colliery.'

Moody and Tyers were discharged from service. Other than having one leg a bit shorter than the other, Moody was to suffer no long-term effects. After a year in Durham county hospital, Tyers came out to a lifetime of pain. 'The pelvis healed but never the internal injuries. I've never been able to travel far because of urinary problems, never more than a few miles. And since that day I've never had a full night's sleep. Being a Bevin Boy wrecked my life.'

It was sobering for Bevin Boys to see or hear about any miner being badly injured, or even killed. Most served in pits that were deemed 'safe', which had few if any serious mishaps. Others had a different experience. Donald Whittle at Newtown near Manchester remembers 'five or six deaths and as many serious injuries'. Les Thomas at Yorkshire Main remembers a terrible accident to a deputy in our area, a nice man with children who'd been ill and probably returned to work too soon. He must have collapsed across the rails where the paddy train ran to and from the face continuously. He was dragged back and forth for a whole shift. His body was so badly mutilated he was brought out in bags.

Jack Garland saw his share of nasty accidents down Markham pit in south Wales, but was most upset by seeing 'a young lad crushed in a fall that broke his pelvis. He was only fourteen, one of triplets who were all down the pit like their father. The miners used to joke he was only a third of a lad he was so small. He came back, but it was a long time after.'

What Peter Rainbow found truly shocking was that when the bodies of three men who'd died in a fall after props were removed on the coalface down the Leycett pit in Stoke 'was that there wasn't a mark on them – it didn't seem they could be dead'. What has stayed in Geoffrey Mockford's mind is that the faces of the two or three colliers who were badly injured in his time at Broadworth Main, Doncaster, 'were white beneath the black surface of coal dust on them. You weren't aware of the whiteness of a man's face normally. Why were you then?' To him, the way in which the injured men were taken from the pit, 'on the conveyor belt with the coal and then on a tub', seemed almost inhuman. For Ronald Griffin in Aberbaiden, that men who'd been hurt 'were just carted away and we carried on working like it didn't matter,' was truly shocking. He felt so badly when he saw his first injured man 'that I went to see his wife, though I didn't even know him'. Alan Munford, who before call-up worked on his father's farm in Kingscliffe, south of Peterborough, was used to the outdoors and found life underground hard to adjust to, but his compassion as a committed Christian drove him to join the mine rescue team at Broadworth Main, Doncaster. He saw a number of men badly injured and one killed in a rock fall 'but it never shook my faith in God. No, I had a very strong foundation of faith.'

The universal pre-war custom when a miner was killed underground was to close the pit for twenty-four hours. Few pits seem to have done that during the war.

As with everything to do with the collieries when they were in private hands, the provision made to deal with injuries ranged from the good to the primitive, and some places barely conformed to the law – which was for every deputy to be trained in first aid and to carry a first-aid box. One County Durham pit went far beyond the requirement, issuing every man working on haulage with a small first-aid kit sealed against deterioration; at the other end of the scale, when John Wiffen in Wales jammed a finger under a tub wheel ('and blood poured out of both sides like juice from a lemon') he had to wait for the fireman to attend to him – the fireman was the only one with a first-aid box.

The best pits had refuge holes along the roadways fitted out as medical stores equipped with rubber tourniquets, splints, canvas stretchers, even stretcher trolleys, a treatment room below ground and a well-equipped ambulance room on the surface, with a full-time attendant, and a vehicle on standby in case of a hospital emergency. The worst pits had nothing more than an additional first-aid box at pit bottom and no medical facilities down below or up on top, and on occasion not even an available van – at one colliery near Stoke, a Bevin Boy reported, a man hurt on night shift was taken home in a wheelbarrow because there wasn't a driver available.

What could be done for injured men underground in this period was very limited; when Derek Agnew had a look inside one medical kit, all it contained were three small ampoules of iodine, three packets of burn dressing and nine of different sized wound dressing. And yet, he noted, 'Chislet was average.' Many smaller injuries treated underground healed well enough, but given the unhygienic conditions there were complications for some, including Ken da Costa, hurt in a freak accident when he was walking down the footrill [drift entrance] and a gearhead motor on the surface broke away and came hurtling down behind him. He didn't know what it was,

> just this thing smashing against the rock, side to side, metal screeching, creating sparks. I got into a safety hole, but chunks of rock hit me in the face, both sides. I staggered to pit bottom and the bloke who did first aid tried to clean me up. But his hands were smothered in oil and coal and both sides went septic.

Da Costa was sent home to receive treatment as an outpatient at a skin hospital in London, having sunray treatment every day to burn off the skin ('mad, but that's what they did in those days. I was left looking like I'd had bad acne'). Six months later a Harley Street specialist's report gave him a discharge from mining.

While there was never any doubting that officials underground did their best for a man who'd been injured, what horrified Derek Agnew was that, if he was able to walk, he was expected to make his own way to pit bottom, often over several miles. It seemed obvious to Agnew that a man might faint and that the danger that he'd be further injured – or worse – by haulage gear or tubs was very real. Geoff Rosling had to go 2 miles, with his 'unreliable lamp, all the time avoiding full and empty drams whizzing past while I took cover in the refuge holes', his pain increasing and fighting waves of nausea; Len Ungate, a print apprentice from Hampton in Middlesex, split a finger to the bone between a tub and a post down the Penallta pit near Caerphilly and bleeding profusely staggered the same kind of distance alone. Neither man's ordeal was over: each had to wait for a lull in activity before he could go to the surface – winding coal took precedence unless a man was very badly hurt. Sent home for treatment, Rosling was gone for three weeks, Ungate for nearer three months.

Roy Doorbar shrugs about the lack of on-site medical facilities. 'It was the dark ages,' he says. 'Men weren't regarded as valuable. At least after nationalisation the NCB [National Coal Board] showed more compassion.'

Accidents aside, pit life was injurious to health. Bevin Boys didn't serve long enough to develop the killer diseases of silicosis and pneumoconiosis that petrified the lungs, or the dreadful eye disease nystagmus,[12] or the chronic inflammatory conditions of beat hand or beat knee (bursitis), caused by overworking the joints in awkward positions. But a number picked up others' illnesses that were common among miners, including pleurisy and pneumonia, caused by the sharp changes in temperature between the pit and the surface, and from working in water.

Buried in a roof fall while ripping in the Florence pit in Stoke in the week before Christmas 1944, Ron Bown developed pneumonia and was taken into the North Staffs Royal infirmary. Nobody thought to inform his family, 'so when I didn't turn up at home for the Christmas break my sister and sister-in-law came looking for me'. In January 1945, in what was meant to be his last week's training at Oakdale in south Wales, Warwick Taylor developed double pneumonia. 'I thought I had flu,' he says. 'I staggered out the billet at the hostel to the sick bay, which meant going out into the pouring rain. "I'm ill," I said – and collapsed in a heap on the floor.' Over the next five days his condition worsened and when he started haemorrhaging from the lungs he was taken 40 miles to hospital in Newport by taxi – which made a detour to a village to pick up a woman in an advanced stage of labour. Taylor was unconscious when he got to the hospital. He wasn't expected to live and his parents were sent for. Luckily, penicillin, a new drug at the time, saved him. When finally he was sent home for six weeks' recuperation it was with 'the souvenir of the two needles they'd used to inject me 56 times'.

Both Bown and Taylor returned to the pits. Unusually, Taylor's medical panel first sent him for three months to work on a gun construction site at Mill Hill because they thought he needed fresh air. Here he helped with electrical installation, 'my hair fell out because of the massive amounts of penicillin I'd been given, and I lost my girlfriend'. Then it was back to Oakdale to do his training all over again.

A variety of physical conditions did get men discharged from Bevin Boy service including Eric Bartholomew, who came back to the Richard Bridge colliery in Accrington after a weekend away, with a doctor's certificate saying he had chest pains. The colliery thought he was malingering; a navy medical panel found that in his eleven months underground he'd developed a heart condition – from which, in middle age, he was to die, by then known as Eric Morecambe. After years of hewing that for some meant lying in confined spaces, many colliers developed arthritis; Syd Walker, always a haulage hand and never a collier, was badly affected by it, young man though he was. After two years his shoulder joints began to seize up, eventually becoming so bad 'I could no longer dress myself and walked about with my hands resting in my coat pockets to ease the pain.' Following six months' hospital treatment as a day patient he was graded C3 and demobbed. His Bevin Boy experience had been fairly traumatic: nine accidents (including twice being trapped when the roof fell in and once when 'a steel girder snapped with a sound like a bomb and the floor began to rise, a gaping crack appeared between my feet and I was about to be squashed between the floor and the roof when the movement stopped with three feet to spare'); and, as he was to find out in later years, the number of times he'd knocked himself down running out at loose it had given him impacted bones in his neck. Repeatedly bending in one direction to release clips gave Gerald Carey bad back pains that an X-ray at the Royal National Orthopaedic Hospital revealed to be curvature of the spine. Roland Garratt suddenly found he couldn't see properly underground, 'everything appeared to look like a dark photographic negative'; the change in atmospheric pressure as the cage went up and down the shaft had caused ear damage affecting the optic nerves. Whether Peter Allen was bitten by fleas in his lodgings or was allergic to coal dust he never knew, 'but I had two attacks of serious blood poisoning and the quack decided I mightn't survive a third'. Stomach trouble saw the exit of Shaffer twin Peter (ulcer) and Dennis Faulkner with what a local doctor diagnosed as gastritis. 'I was a very fit young man, swimming and gymnastics, but given to severe attacks,' he says. 'Conditions down the pit exacerbated things: I kept vomiting and being signed off and on. A medical tribunal in Pontypridd found me unfit to work in a coal mine – but it took 40 years before I found the real cause.'[13] The polio epidemic of the hot summer of 1947[14] claimed Tony Brown and Reg Taylor. 'On Friday afternoon I was in the pit; on Monday morning I was in Lodge Moor isolation hospital,' remembers Brown. It was six months before he was able to go back to work – in his trade as a signwriter. Taylor was on the danger list and in hospital until Christmas and it was another six months before he was fit for employment – again as a civilian. He was left with a weak arm and leg, Brown with two weak legs. 'But I made a 70–80 per cent recovery and that was okay,' Brown comments. 'Another chap at the hostel and the son of the night watchman got it worse, in the lungs.'

During a debate in the House in December 1943, Lieutenant Colonel Lancaster raised the matter of whether conscript miners should be psychologically tested. Ernest Bevin turned that down as being impractical. But some men had a rough time of it mentally. Jack Garland knew he was claustrophobic (he'd found out on holiday at Newquay when he went into the caves with his father and brother), was ashamed about it and said nothing, 'but I was never happy, never comfortable down there.' Others tried to cope and failed, like Dan Duhig and John Marshall, who both suffered nervous breakdowns. 'I had no reason to think I wouldn't be okay,' says Duhig.

I didn't like being shut in – maybe I was shut in a cupboard when I was small or something like that – but I didn't see it as a problem. As a boy I was the one who jumped off a shed roof thinking I'd float down like thistledown, and hit the ground like a sack of spuds. I was the one who stood on one end of the seesaw when the others jumped on the other end to see what would happen to me. They asked me at my medical if I was scared of blood and I wasn't, I'd pulled bodies out in London during the Blitz.

But he never got used to 'the dark, the closed-in feeling', down the Stobswood pit near Alnwick in Northumberland. He stuck it for over two years – and then went to pieces.

I can't tell you how it came on. I don't remember anything, just being spread-eagled on the bed that I shared with the landlady's son. I have no memory of selling my clothing coupons to get home, which later I found out I did, or of the medical that made me grade 4 and got me out. A whole part of my memory had gone. I could remember early childhood but I'd lost things that happened after. I loved mental arithmetic but my ability completely disappeared. Years later, when my wife was alive, I'd go up to the spare room with a piece of paper and a calculator – and nothing. I joined a discussion group and when it was my turn to speak I'd know what I wanted to say, go up the front – gone. A psychiatrist at the John Radcliffe hospital in Oxford advised me to read poetry and I learnt Kipling off by heart and would then remember not a thing.

Posted to the Stargate colliery near Gateshead, John Marshall 'felt a great heaviness around me. The darkness frightened me, it clung to me.' And he couldn't rid himself of the fear that he would be made to remain in it 'for ever and ever'. When he told his colliery manager he was claustrophobic, 'he just thought I was trying to pull the wool' and had him moved to another pit, Greenside near Newcastle. Gradually things got on top of him:

I couldn't sleep at the hostel with people on different shifts coming and going, and the noise in the pit gave me a terrible pain in the head. Sound was my business, I had sensitive hearing, and the noise in the pit was killing me – my hearing was being affected by explosions.

I was sent to work with an old miner who had one eye, I can see the sod now, shovelling his coal on to the belt. I shovelled and shovelled but his pile grew and grew. And all he kept saying to me was 'Shovel'. Eventually I could stand no more and I threw my shovel straight in his face. The mine manager said I was a troublemaker but I told him, I'd never been that. I was obsessed in wanting to know who had drawn my number, who was responsible for me being down the pit. I told him, 'I have a right to know that and no one will tell me.' I was fined £10. I had trouble paying – I borrowed off the other lads.

I was heading for a big breakdown and I had a very bad experience down the pit. I was so desperately tired that when I got a chance to sit down I fell asleep, for some reason having turned off my lamp, a ridiculous thing to do. And when the shift went up I was left behind in a cubbyhole. When I woke up there were a lot of tiny green lights going on and off around me. It was the rats, blinking, which scared the hell out of me. I tried to relight my lamp. What I should have done was stay where I was and wait for someone to come and find me, but I had to get out, so I followed the line of the tracks. Suddenly I found they were rusty so I was going the wrong way – I'd arrived in the old workings. And the smell in there, rust and damp and foul air: I went off my head.

A rescue party did come looking for Marshall. At the hostel the nurse put him to bed, but he ran away, going back to London but not going home. He rang his mother later and she told him the police had come looking for him. His recollection of the period is hazy.

I was in a bad way, going round in a daze, getting worse and worse. Then I got a septic throat caused by the coal dust and was taken into hospital, the big one in Smithfield, St Bartholomew's, for an operation. I was in there a month and then they said, 'You can go back now.' 'No, that's it,' I said, 'I'm not going back, you can do what you want with me.' Eventually they sent me to see a psychiatrist who examined me and gave me tests and said, 'They should never have sent you to a coalmine.'

Whether any Bevin Boy became so unstable as to take his own life has never been established, but Brian Folkes relates:

When I was home on leave on one occasion I learnt that the nephew of two elderly ladies living opposite had shot himself rather than return to his pit in the Midlands. I have no hard evidence, but these two highly respectable ladies told my parents and I see no reason to disbelieve it.

There is no record of how many Bevin Boys were killed working in the pits. The newspapers reported on the death of the first: an unnamed conscript who in March 1944 looked up the shaft at the Heworth pit near Newcastle and was struck by the descending cage; and what may have been the death of the last: Gerry Moore and Eric Martin in August 1947, in an explosion in the Louisa pit 5 miles away, which also claimed nineteen regular miners.

Anecdotal evidence from other Bevin Boys records other deaths, with County Durham figuring in another two of them. Peter Allen (Ryhope, Sunderland):

> A Bevin Boy was at the bottom of the shaft after midnight and placing the tubs in the cage. When the tubs were in place you were supposed to ring the bell so the engine windingman knew he could haul them to the surface. On this occasion the Bevin Boy rang the bell, then realised he hadn't secured the tubs properly. Instead of telephoning to the top first, he went into the cage to attach them at the moment the cage was hoisted. He was caught half in and half out and died instantly.

Des Knipe (Derwent, Newcastle):

> A set of tubs shot the rails and got away at Eden, the next colliery to mine, with a Bevin Boy hanging on the tail rope, something no one was supposed to do but we all did. He was crushed. I don't remember his name, a Durham lad, Victor or Vickers, but I knew him, we were in the hostel together and we had a friendship. It happened.

Donald Whittle (Swinton, Manchester):

> I stayed on at Swinton after training and a Bevin Boy was killed in the training section about a year after I'd finished there. A runaway tub struck a group of Bevin Boys at the bottom of a gradient. I kept a diary through the period. The entry of 20 October 1944 reads: 'Bad accident in the trainee centre – 1 killed and 4 hurt badly'.

Les Thomas (Yorkshire Main, Doncaster):

> A Bevin Boy and a regular young miner died because a 14-year-old who was operating the squeezers felt unwell. There was a bit of a decline around a bend to pit bottom and there were 'squeezers' installed that brought the tubs to a final halt before the cage. This little lad went off for a bit of air and left the squeezers open. A truck filled with stone couldn't take the bend and piled through and landed on top of the pair of them. They died on the surface. They changed the rules, said no lad so young should be in charge of the squeezers in future, but it was too late then, wasn't it?

With passing references in newspaper reports to the death of a Bevin Boy in Wombwell Main near Barnsley in a similar cage accident to that related by Peter Allen; to another, in a roof fall, in Markham pit near Caerphilly; and to two more caused by runaway tubs in Cannock Chase and 'a Midlands pit', the total is eleven. It may have been more.[15]

John Cook would have been among the number, whatever it was, had he turned in for the night shift at the Louisa pit with his friend Gerry Moore, whom he'd met on his first day as a trainee and who'd recently been his best man.

I was detailed for duty with Gerry, but another friend, George Warriner, who was older than me, was being demobbed, and after working that last morning with us was leaving from Newcastle that evening. I was due to go down at midnight on a cable-laying shift, but I went to Newcastle to see him off and we had a few beers. I returned to the hostel in time for work and got changed. But I felt so tired, not up to a heavy night's shift, that I changed back and went to bed.

Six

Getting to Know You

There was prejudice on both sides to begin with. Most Bevin Boys expected all miners to be uncouth and hostile. Miners expected Bevin Boys to prove incapable of the job, especially the conscripts who, to a man, they assumed, came from privileged backgrounds. Each side treated the other warily, but with curiosity.

For their part, the Bevin Boys were relieved to find the majority of the men in whose midst they found themselves 'very accepting and warm-hearted', as Harold Gibson puts it. 'I never found any antagonism towards us,' says Denys Owen, an audit clerk in a Liverpool chartered accountants. 'They possibly were amused at us innocents being sent to join them, but I think they were glad of the help.' 'Miners were just very decent people,' says Alan Gregory. 'But there might have been one other factor that strengthened their goodwill. Miners had for generations felt that they were hard done by. The fact that college boys and suchlike had been directed to work underground somehow elevated their status.'

In the small Yorkshire Horse Riggs pit, David Reekie found the forty or so miners there more than accepting. 'They looked after me,' he says. 'As the only Bevin Boy in the place I was some sort of mascot in a ridiculous way.' Even the aloof Anthony Shaffer found the miners 'helpful and sympathetic', though his memoir adds with irony typical of him: 'They didn't, curiously enough, resent us as they might have, seeing that from their point of view we were toffs temporarily forced to share their misery.[1]

Some miners did resent the strangers the government imposed on them and a few Bevin Boys had a tough time of it initially. That was particularly the case for the German Victor Simons at Bolsover in Derbyshire.

When I first started there were comments, 'Buggered if I'll work with a fucking German.' A lot of that kind of thing. I understood they saw me as an exotic creature: a member of the enemy nation, a public schoolboy, very dark complexioned – I was naturally dark but after nine months on a farm I was very dark from the sun, where they were so white beneath the pit dirt.

Their attitude was to be expected. But I did not pretend to be other than an ordinary person, though admittedly somebody ambitious to work his way up in the mining industry. Eventually I was accepted because I mucked in.

So, too, were Meir Weiss and the other *kibbutzniks* at the Ashton Moss colliery near Manchester, and for the same reason. 'We met no antagonism but they couldn't make us out,' he says.

> Our English had much improved since we had arrived in this country and we always spoke English together, not German, which was sensible and polite. But it was complicated to explain how and why we were there. We were a great puzzlement to the miners. They thought we must be gypsies. Possibly they had seen Jews in the local market. In their belief Jews did not work. We worked hard.

Bevin Boys who did work hard was acknowledged by the miners and treated as one of themselves; a few were even appointed as workforce representatives on their pit production committees. 'Really, the locals and the Bevin Boys – there were 20 or 30 of us in the Homer at Stoke – were just blokes in an integrated company of men,' Doug Ayres says. Peter Archer from Wednesbury in Staffordshire, who'd been marking time to call-up as clerk to the Birmingham district auditor, agrees. 'I don't think Bevin Boys and regular miners were distinguishable eventually. We were in this together. I enjoyed the feeling that we were a team, getting the coal up the shaft even if the conveyor belts were held together with elastic bands.' In their hearts, however, most miners must have considered the Bevin Boys as birds of passage; as Big Joe earnestly told Geoff Baker: 'They can't joost become a miner, Joffers. They must be born to it.'

Bevin Boys found the miners to be men of contradictions. Many were religious, stalwarts of their church or chapel, but were equally blasphemous. At Norton colliery at Smallthrone near Stoke, the fireman in charge of the seam where Roy Doorbar worked was a lay preacher,

> and when the full tubs were coming up the dip and going to pit bottom he'd get on his knees and pray out loud, thanking the Lord for all the coal that was on its way. But if any of the tubs came off the rails and the coal went everywhere he'd stand and shout, effing and blinding. The incongruity used to amuse me no end.

It amused Bevin Boys generally how superstitious many miners were. Some older ones frowned on whistling in the pit, believing it could summon the Devil. An almost universal superstition was that seeing a hearse or an ambulance was an ill omen. Walking from his hostel to his pit in County Durham, Roger Spencer once 'met a miner on the road who was so worried when we were overtaken by an ambulance that he immediately turned and went home, convinced a disaster would occur at the pit. I went on to work and had an uneventful shift.' Miners everywhere were reluctant to work on a New Year's Day, believing it the unluckiest day in the calendar, and in Staffordshire, certainly, none did after fifty-seven died in an explosion at Berry Hill, Stoke, on New Year's Day 1942.

When disaster struck, when comrades were trapped or badly injured, miners thought nothing of risking their lives for them yet, inexplicably, some would steal another's tools without compunction. 'The camaraderie was tremendous, but if someone had a better pick, they'd take it,' comments Alan Gregory. Some collieries had racks at pit bottom to lock up tools, but most men preferred to hide theirs where they were working – it was an awkward business carrying picks and shovels through low roadways, possibly for miles. But if another miner came across tools and they weren't stamped with the owner's name, they were fair game. 'It has to be said,' adds Gregory, 'that some men were much less skilful at hiding their tools than others.'

Miners were not, of course, all of a kind. Some were serious men who served as local councillors. Some couldn't write and had difficulty reading, but others were widely read or knowledgeable about classical music. A collier at Chislet in Kent often sat beside Derek Agnew at snap 'quoting reams of Kipling'. At first he thought this a 'fantastic sight' but decided he was quite wrong; why shouldn't a miner be a poetry lover?[2] 'You can find culture in the most unlikely places,' says Norman Brickell.

Taking tubs around the pit I used to stop for a chat with an old man affectionately known as 'Uncle Wire' Horner, who had a little job sweeping up, whiling away his time before retirement. I liked to be there around snap time because he shared his pork sandwiches – he kept a pig or two. One day he asked did I know that for the first time since the end of the war Gigli, the famous Italian tenor, was singing at Covent Garden and was being broadcast that night, and urged me not to miss hearing his voice. Picture this dear old man, covered in coal dust, ecstatic about being able to hear Gigli on the radio.

And while some miners were teetotal, the majority were heavy drinkers who indulged in weekend sprees and gambled heavily into the bargain: horses, dogs, anything. 'On the bus ride after a shift the miners would bet on whatever came into their heads,' observes Geoffrey Mockford. 'They made bets on events which could take place on the journey. A favourite item was bus ticket numbers: the highest total of all the figures, the highest total of a particular digit, or a sequence of numbers. Anything. Even flies or raindrops on the window.' Ernest Noble, a clerk on the London Midland & Scottish Railway in his home town of Bradford in Yorkshire before becoming a Bevin Boy, got into an argument with a fellow conscript down the pit at Frickley in South Elmsall which the miners decided could only be resolved in the boxing ring – so they could bet on the outcome. 'Both of us were reluctant and each of us thought he'd won, but the overman who refereed decided it was a draw, which was brave of him seeing how much money was put down.'

It took a while for miners and Bevin Boys to reach some kind of understanding, but they generally did. And Bevin Boys came to respect the miners for their resilience (in an era of no work no pay, it wasn't unusual to see a miner who'd been hurt but not incapacitated carrying on working) and to like them for their cheerfulness. Down the pit, miners were happy. 'Are you happy, Jimmy?' was asked so often of Jim Bates that he concluded 'being able to be happy in adversity was probably the most important thing in life'. Perhaps, above all, Bevin Boys ultimately liked the miners for the generosity of spirit that lay beneath their often rough exterior, but Owen Jones took longer than most to appreciate the miners' less obvious traits.

> I was brought up in a home, a Puritan upbringing, and I was very sensitive. I suffered at the miners' hands at first. Miners were hard and uncompromising men and they didn't hold back in telling you what they thought of you if you weren't doing your job to their satisfaction. But I learnt to do my job, and I found out that most miners were good men. I have no complaints. Being a Bevin Boy was a crash course in helping me loosen up as a person and learn to stand up for myself.

'A lot of miners were just brilliant blokes,' says Geoff Darby.

> One was Abe, who had a fantastic voice, like Bing Crosby, who knew all the songs and once he started up he'd have the whole face singing. He'd finish his stint and he'd help another face miner. He helped me many a time when I was tramming [putting] for him. 'You rest a minute,' he'd say and finish the shovelling.

'Older miners had a lovely attitude and some of us formed strong friendships,' says David Roland. 'It was a different matter with the local pit lads, who really were just boys, only 14 or 15. We had some trouble with them. They resented us, especially ones like me with London accents. They especially resented those of us who were asked to work at the coalface. Understandable, really.' Adds Geoffrey Mockford:

> I don't think Bevin Boys noticed the pits lads. We had nothing in common, nothing to say to each other. I couldn't understand what they were saying, anyway, the north Derbyshire accent was so strong. They seemed so small,

The Evening News cartoon published on 7 January 1944 took a light-hearted view of the Bevin Boys' fate. *Boy's Own* magazine presented an idealised view of mining and how much fun the Bevin Boys were having. And the posed publicity photos of the initial intakes when they arrived at the training centres implied that the Bevin Boys were enthusiastic about the part they were to play in the war effort.

A cage raised coal at 70 feet a second and dropped men at 30 feet a second, the engineer in the winding house *(bottom)* priding himself on being able to come imperceptibly to a halt. Trainee Bevin Boys *(below)* were often given an 'initiation drop' – at coal speed.

Many of the faces in the photos, such as that taken at Oakdale in Wales *(opposite)* or Cramlington Lamb in Northumberland *(above)* betrayed what most Bevin Boys felt. Like other 18-year-olds, they had expected to fight for their country, not go down the pits. Some, like Warwick Taylor *(circled, opposite)* served until the war was over and then refused to carry on, winning release into the armed forces. The majority stuck to it, some for over four years, before being discharged.

Oakdale colliery – an archetypal pithead *(above)*. Down below, most Bevin Boys worked on haulage, marshalling the tubs, pulled by cables along rail tracks. In some pits, Bevin Boys used ponies – and in a few, just muscle power *(bottom right)*.

Other Bevin Boy jobs included helping the shot firer, who fractured the coalface for the colliers to bring down the coal, in some pits by hand-drilling the holes for the charges, in others using compressed air drills. A Bevin Boy *(top, opposite)* rams in the shot after drilling. Only a few Bevin Boys became colliers themselves – the toughest job in the pit: tougher still in those pits not yet mechanised *(bottom, opposite)*.

The dirtiest job in the pit was reckoned to be the loader end, where the coal tumbled off the conveyor belts into the tubs *(centre, right)*.

A number of Bevin Boys drew what they saw. At Horse Riggs colliery near Leeds, David Reekie hauled the coal dug out by a miner working in a seam down to 14 inches *(right)*; Jim Bates sketched the miners knocking off at the end of a shift at Hem Heath, Stoke-on-Trent *(centre, left)*; and Dewi Bowen a pit rescue at Elliot colliery in New Tredegar, south Wales *(bottom)*. 'Driving' one of the stationary engines that operated the haulage cables was considered a cushy job – as an unknown Bevin Boy rather exaggerates *(centre, right)*.

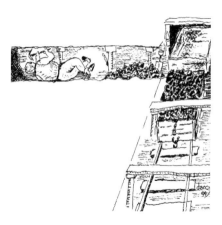

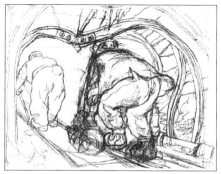

Pony and putter – capturing the close relationship between man and pony *(above)*. *Left:* In a low seam the tub very often hit the timbers of the roof – and getting through was exhausting work. *Below:* Roof fall – a constant danger in every pit.

A Bevin Boy at Craghead, County Durham, Ted Holloway returned to mining five years after he left, later qualifying as an art teacher and teaching in Jarrow. He became a full-time artist in 1981. He was working on a large series of pencil and graphite drawings of pit life when he died in 1987. The three here are produced by permission of his widow Gill.

Going down: a clean-faced Bevin Boy calls into Stores before his shift. *Below:* Black at 'loose it' – the end of the shift – at Askern Main in Yorkshire. The tall figure third from left is John Platts-Mills, a barrister who moved in Churchill's circle and who volunteered for the pits.

Arthur Gilbert *(on the right)* fell on his feet when he became the safety officer at Berry Hill pit, Stoke. The job was sometimes particularly dangerous – but he was his own boss.

Bevin Boys felt sorry for the pit lads, regular miners at 14, who underwent the same training. The four-year age difference was marked, as the group photo of Bevin Boys above shows. *Left:* A pit lad proudly displays his 'lighthouse' safety lamp.

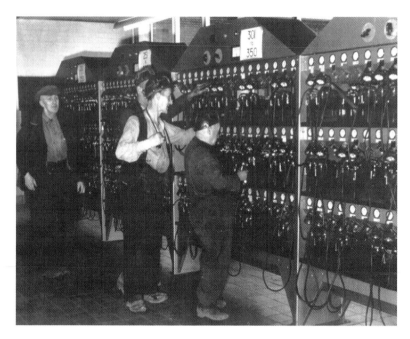

Electric cap lamps began to replace wet-cell 'lighthouses' in the 1920s, but most pits still had the latter during the war. Both types were recharged in a colliery's lamp house *(top)*. Cap lamps were reliable, lighthouses weren't. Many were too old to be maintained properly and often went out underground. Officials still carried oil safety lamps *(above)* because the flame indicated the presence of methane.

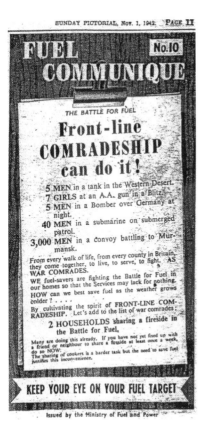

FUEL COMMUNIQUE No. 10

THE BATTLE FOR FUEL

Front-line
COMRADESHIP
can do it!

5 MEN in a tank in the Western Desert.
7 GIRLS at an A.A. gun in a Blitz.
5 MEN in a Bomber over Germany at night.
40 MEN in a submarine on submerged patrol.
3,000 MEN in a convoy battling to Murmansk.

From every walk of life, from every county in Britain, they come together, to live, to serve, to fight. AS WAR COMRADES.

WE fuel-savers are fighting the Battle for Fuel in our homes so that the Services may lack for nothing. HOW can we best save fuel as the weather grows colder? . . . .

By cultivating the spirit of FRONT-LINE COMRADESHIP. Let's add to the list of war comrades:

2 HOUSEHOLDS sharing a fireside in the Battle for Fuel.

Many are doing this already. If you have not yet fixed up with a friend or neighbour to share a fireside at least once a week, do so NOW.
The sharing of cookers is a harder task but the need to save fuel justifies this inconvenience.

KEEP YOUR EYE ON YOUR FUEL TARGET

Issued by the Ministry of Fuel and Power

Government adverts constantly urged householders to use coal sparingly. The public responded. Even so, coal production fell below the demands of war, and more miners were needed. That some conscripts would be sent to the pits was on the cards as early as summer 1942. The announcement came on 3 December 1943 – as *The Times* reported.

YOUNG MEN FOR
THE MINES

COMPULSION AFTER A
BALLOT

MR. BEVIN'S STATEMENT

From Our Parliamentary Correspondent

Because voluntary methods of recruitment have failed to produce the number of new entrants needed to maintain the man-power of the coal industry the Government have decided that some thousands of men between 18 and 25 who would otherwise go to the services must now be " directed " to the mines.

Mr. Bevin was confronted with the difficulty of deciding how these young men for the mines should be selected. As he informed the House of Commons yesterday, he has decided that the only fair method is to make the choice by ballot. A draw will be made from time to time of one or more of the figures from 0 to 9, and those men eligible whose National Service registration certificates happen to

Ernest Bevin, Minister of Labour and National Service in Churchill's government, who sent one conscript in 10 down the pits. Those who went, it was said, had their numbers drawn from Bevin's hat by his secretary.

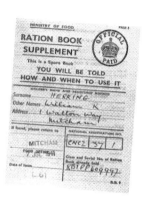

MINISTRY OF FOOD

FREE!

RATION BOOK SUPPLEMENT

OFFICIAL PAID

This is a Spare Book

YOU WILL BE TOLD HOW AND WHEN TO USE IT

HOLDER'S NAME AND REGISTERED ADDRESS

Surname HERRING

Other Names William R

Address 1 Walton Way Mitcham

If found, please return to

MITCHAM
FOOD OFFICE

NATIONAL REGISTRATION NO.

ENC2 37 1

Class and Serial No. of Ration already held

Date of Issue

KB1PF609997

SUPPLEMENTARY CLOTHING COUPON SHEET SC5G

UH 990893

Name HENRY GEORGE PERKIN
(BLOCK LETTERS)

Address School House Abbots Bromley
(BLOCK LETTERS)

STAFFS.

Nat. Reg'n (Identity Card) No. DTXI 141 1 6

IMPORTANT.—These coupons may not be used unless the holder's name, full postal address, and National Registration (Identity Card) Number have been plainly written above IN INK.

CLOTHING COUPON (×16)

MINISTRY OF LABOUR AND NATIONAL SERVICE.

London + S.E. Regional Office, (R 0 5)
HANWAY HOUSE,
RED LION SQUARE, W.C.1.

29.12.43 :(Date)

Dear Sir,

I have to refer to the notification which has already been sent to you informing you that you had been selected by ballot for underground coalmining employment. I have now to notify you that it is proposed to direct you to attend in the near future at the North Staffordshire Emergency Training Centre at Stoke-on-Trent for a course of training with a view to subsequent employment in coalmining.

I enclose a leaflet E.D.L.94 which gives information about training for and employment in coalmining.

You may appeal against this notification if you consider that there are any special circumstances connected with coalmining which would make it an exceptional hardship for you to be employed on that work. I have to remind you, however, that 'at the time of your medical examination under the National Service Acts you had an opportunity to apply for postponement of liability to be called up under these Acts. Accordingly, if you appeal against this notification, your appeal should show in what way you consider that employment in coalmining would be an exceptional hardship to you having regard to the fact that either you made no application for postponement of call up or your application has been determined and postponement, if granted, has expired. If any new facts have arisen since you previously had an opportunity of applying for postponement, or renewal of postponement you should call particular attention to them in your appeal.

If you should desire to make an appeal you should do so in writing within 4 days and address it to this Office.

If you make an appeal it will be put before the Local Appeal Board and the Board will make a recommendation which will be taken into account.

Unless it is decided after appeal to a Local Appeal Board that you should not be directed to coalmining employment you will be required to attend at the Training Centre.

On completion of training at the Training Centre you will be posted to a working colliery and any preference for employment in a particular area which you may have expressed in reply to the earlier notification which was sent to you will then be taken into account as far as possible.

Yours faithfully,

for Regional Controller.

Mr. G.C.L. Baker.

E.D.698   (3295) Wt. 42729/3156 50m 12/43 C.&Co.   745(B)

MINISTRY OF LABOUR AND NATIONAL SERVICE

Emergency Powers (Defence) Acts, 1939-1941

DIRECTION ISSUED UNDER REGULATION 58A OF THE DEFENCE (GENERAL) REGULATIONS, 1939.

NOTE.—Any person failing to comply with a direction under Regulation 58A of the Defence (General) Regulations, 1939, is liable on summary conviction to imprisonment for a term not exceeding three months, or to a fine not exceeding £100 or to both such imprisonment and such fine. Any person failing to comply after such a conviction is liable on a further conviction to a fine not exceeding five pounds for every day on which the failure continues.

To Mr. G.L.L. Baker
St Lukes Cottage
Victoria Docks
E. 16. (Date) 17 JAN 1944

FREEMASONS ROAD,
E. 16.

In pursuance of Regulation 58A of the Defence (General) Regulations, 1939, I, the undersigned, a National Service Officer within the meaning of the said Regulations, do hereby direct you to perform the services specified by the Schedule hereto (see overleaf) being services which, in my opinion, you are capable of performing.

If you become subject to the provisions of an Essential Work Order in the employment specified in the Schedule, the direction will cease to have effect and your right to leave the employment will be determined under that Order. Otherwise, this direction continues in force until 23. 7. 44 or until withdrawn by a National Service Officer.

I hereby withdraw all directions previously issued to you under Regulation 58A of the said Regulations and still in force.

National Service Officer.

E.D. 383A.

[P.T.O.

(3314) Wt. 6041/3040 250m 4/43 D.&Co. 745(B)

Above: The notification received by Geoff Baker, informing him he was to be a Bevin Boy. *Left:* The threat of prosecution and imprisonment that accompanied the notification. Some conscripts chose prison.

There were small compensations in being a Bevin Boy: extra cheese and extra clothing coupons – but not the money to make much use of the latter.

It was impossible for Bevin Boys from the south to get back on time after Christmas. Hundreds chose not to, like Dave Moody, and were prosecuted and fined.

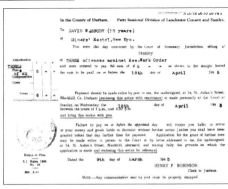

An apparent threat from the union at Bullcroft colliery in Yorkshire brought the Bevin Boys there out on strike – they were shipped off to the army. *Right:* Servicemen returning after the war had the right to reinstatement in their jobs. Bevin Boys, like Harold Gibson, hadn't.

Some Bevin Boy hostels were purpose built; others were old refurbished army transit sites. All, like Pelaw Bank in Scotland *(top)* and Oakdale *(inset)* in Wales, were basic, 'ovens in summer, refrigerators in winter'.

Below: A typical hut interior, photographed at Abbotts Road hostel in Mansfield, Nottinghamshire, by Geoffrey Mockford.

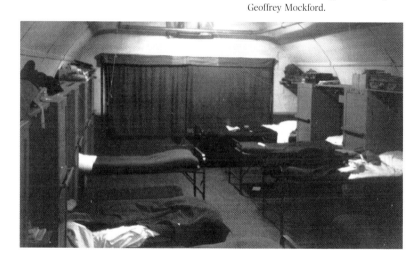

The hostels provided a reading room – Geoffrey Mockford also photographed this one in Mansfield – where Bevin Boys could study for their future, if they wished. After a hard shift, most were too tired.

Not all Bevin Boys were in hostels, where the inmates had showers. Many were in digs some of which had no indoor sanitation, and a tin bath in front of the kitchen range was the way of life – little different to how things were in the 1930s *(left)*.

No record exists of how many Bevin Boys were killed in the pits. The two last died in the County Durham Louisa pit explosion in August 1947, along with 19 other men. One was Gerry Moore, third from right.

On 1 January 1947, the coalmines were nationalised, the blue and white National Coal Board flag raised at the pitheads. 'We're the bosses now,' the miners said, and believed they would have a say in running their industry. But little really changed and the Bevin Boys watched as disillusion set in.

about half the height of us. The four years age difference was part of the explanation, but they'd had a hard upbringing. Their growth was stunted. I felt sorry for them.

Miners found amusement in all but the most dire of situations and their sense of humour was often, appropriately, black. Doug Ayres remembers an incident in which an injured man was bleeding profusely and tourniquets and various dressings had failed to stem the flow.

> One of the others said, 'I'll fix it,' put his hand in his pocket and pulled out a crumpled strip of paper, which he rolled into a ball and put on the wound. The blood stopped immediately. 'What was that?' his companions asked. 'It's me wage slip,' he said. 'There's enough stoppages on that to stop bloody Niagara Falls.'

Miners found amusement in almost any situation underground and 'took the mickey constantly', Doug Ayres recalls. They were interested in the Bevin Boys' backgrounds and what sort of lives they'd be returning to. A miner called George Meredith one day asked Geoff Baker,

> 'Wot at gooin' ter do wen thee goost wom?' 'I might go in the Church,' I replied. 'Thees moor money in minin' that ther is in vicarin',' George advised me. 'Remuneration isn't the prior consideration, George. One is called to that life,' I said loftily. 'They beyin' cowed!' exclaimed George. 'Ah 'erd they bloody cowed, but it were never a vicar!'

Bevin Boys often found themselves the butt of practical jokes. It was easy to take in almost every bewildered novice but some were more gullible than others. A favourite trick was to send a newcomer for a roadman's hammer (a non-existent tool that could fix anything) or even, as Roy Doorbar relates, 'a bucket of steam for the steam engine. And some of the silly buggers went all the way to pit bottom for it. And the on-setter would send them back saying he hadn't any left.' Some Bevin Boys believed it when they were told that the gurgling noise of subterranean water that could occasionally be heard when a hole was bored in the coalface was being made by a coal frog; or that the surveyor's theodolite was a camera brought down to photograph them, and they waited days for their prints to arrive. At Edwinstowe colliery in Nottinghamshire, Harry Fowler recalls how he and other Bevin Boys were taken in when they were told 'the pie lady came down the pit on Fridays, so we needn't bring any snap. So we didn't, and we waited and waited. Of course, there was no pie lady.' At Oakdale in Wales, Warwick Taylor recalls, 'the regular miners used to enjoy sending new Bevin Boys to work with the ponies as we didn't speak Welsh and the ponies didn't understand English'.

Michael Edmonds still laughs at a couple of pranks played on Bevin Boys down Bedwas pit near Caerphilly.

> A new Bevin Boy arrived at the bottom of the shaft for his first day's work and he was very apprehensive. He was taken past a journey [set of tubs] of waste rock filled on the night shift to a pile of brattice cloth lying among sacks of stone dust. One of the butties nudged him. 'See that cloth, mun, there are dead men under there, killed last night – a journey will be along soon to take them out.' The Bevin Boy paled a bit.
> The other incident involved a collier who'd been having trouble with the Bevin Boy sent to work with him. He got hold of an old pair of mining boots, set them upside down in a pile of rock, and next morning led the Bevin Boy to the fall. 'See that?' he said. 'Well, that's where I buried the last Boy I had with me, just the same as you, full of himself.'

Sometimes the humour was more physical. A Bevin Boy particularly nervous about travelling in the cage might have his wrist wired to it so that he had to endure going up and down until someone took pity on him. Or a loader end would be moved back a little from the track so that a Bevin Boy working there would be unable get his tub under it, however he sweated and strained, scared he'd be taken to task as the coal pouring off the conveyor cascaded around him. Norman Brickell was on the receiving end of a kind of induction that was once common in many trades but to which no other Bevin Boy seems to have been subjected:

> At snap I'd just put the feedbag on Dandy's head when I was suddenly surrounded, my trousers were pulled down around my ankles and I was forced to the ground. The cap of one of the oil drums was removed and its thick black grease was poured over my genitals. Great hilarity all round. But there really wasn't anything nasty about it. It was a way of showing I was accepted. 'Nah then, Norman lad,' they said, 'tha's one of us.'

Bevin Boys also had a sense of humour (those who served at the Welbeck colliery in Nottinghamshire remember a big notice at pit bottom warning IT IS HIGHLY DANGEROUS TO WALK ACROSS THE SHAFT, on which one of their number wrote ESPECIALLY AT THE TOP); and once they'd found their feet they weren't slow in joining in the practical joking. When he became his colliery's safety officer, Arthur Gilbert acquired a powerful hand lamp of the type carried only by three or four officials, and towards loose it, whenever he spotted a group of men 'idling in a spot where they could make a quick getaway to pit bottom,' he'd douse his lighthouse, creep towards them, then highlight them in his hand lamp's beam, making them 'jump to their feet feigning useless employment of some description.

Then the penny would drop and I became the recipient of collective abuse.'
A Bevin Boy in Stan Payne's pit frightened the wits out of a shot firer he was
helping:

> They'd set about a dozen charges, looped together to fire simultaneously,
> retreated to their firing point, and nothing happened when the key was
> turned. Annoyed, the shot firer issued a curt 'Stay there' and went to check
> the connections on the face. On returning to the firing point he was greeted
> with 'They still won't blow.' From the startled shot firer: 'How do you know?'
> From the straight-faced Bevin Boy: 'Because I've tried it twice while you've
> been gone.'

Another memory:

> A Bevin Boy who was an amateur ventriloquist was sent with an elderly
> ripper to clear a small roof fall. While they were working, more rock fell and
> the Bevin Boy switched his light off and threw his voice under the new pile.
> 'Help, help, I'm under here.' So the old chap started clearing the rock away
> as fast as he could go. After several minutes the Bevin Boy threw his voice
> into the tub where the rock was. 'Help, I'm in here now.' The shot firer
> wasn't sure whether he was having his leg pulled or not. The old ripper did.
> He stopped, wiped his brow and said, 'Well, get theesen oot. Tha should ha'
> called oot when tha were on t'shovel.'

Geoff Baker probably trumped anything the miners got up to at the Norton pit in
Staffordshire the day he went down in a dinner suit.

> On a visit home I bought it for sixpence at one of my mother's jumble sales.
> It was a perfectly good dinner suit, but there wasn't a great demand for
> dinner suits in London Dockland in the war and there were no bidders. I
> took it to the pithead baths in a brown paper parcel, changed, and appeared
> out of the dark into the light of the lamphouse, complete with my wide,
> shiny lapels, white shirt and dickie-bow, the outrageous ensemble set off
> by pit helmet and broad leather belt. And down I went, saying I'd been to
> a party the night before and hadn't had time to change. The story set the
> colliery alight.

*   *   *

Oscar Wilde was referring to the Americans, but Bevin Boys would have
substituted the miners: they and the miners were separated by a common
language.

Britain in the 1940s was rich in regional dialects and those Bevin Boys sent to pits outside their home areas at first floundered as they tried to understand what the miners were saying. When the miners slipped into pit slang – 'pitmatic' in County Durham – Bevin Boys were completely lost. As far as John Squibb is concerned, what he listened to down the Shop pit in Gateshead 'might just as well have been a foreign language'. In Wales the added difficulty was that miners often broke into Welsh when speaking among themselves and sometimes addressed Bevin Boys in it – inadvertently or simply because their bewilderment amused them. Posted to the pit in his own Staffordshire village, Roy Doorbar was in a position to enjoy watching his fellow Bevin Boys struggle with 'potsy', the dialect of the Potteries. 'The old miners would talk between one another in real Pottery slang and the poor Bevin Boys hadn't a clue they were being called silly buggers and the white brigade,' he says. 'I used to smile to myself because the miners didn't know I was a Pottery lad and I didn't let on. But a few weeks later they heard my name and realised I was Danny "Duba's" son, who'd been down the pit in the twenties.'

Doug Ayres, who served in the Homer and Sunderland in Stoke, was fascinated by the etymology of posty. 'The local paper, the *Evening Sentinel*, in its Saturday edition, included a paragraph or so set in dialect – in some quarters the dialect was highly prized because its roots clearly were in the 16th century,' he says.

The first person plural of the verb conjugated with the 'm' sound, for example 'we'm go' or 'we maun go' either meaning we must go, and it could represent the use of I am down to we am, in a shortened form. The second person singular was used with the nominative form pronounced 'they' and the accusative and dative pronounced 'thee'. The auxiliary verbs remained as in the old days, for example 'whist they bin?' meant where hast thou been? Or 'artowrate?' meant art thou all right? 'Woot' meant would'st thou'. More obviously 'canna' meant cannot and 'dunna' meant do not. Some special vocabulary from Old English included 'fang ite': take hold of – very important to understand if you're handling steel rings [half-arch girders], which are extremely heavy, shaped like a hockey stick and have a centre of gravity in fresh air.

In their five years in this country, the Standard English acquired by Meir Weiss and his fellow Jewish refugees was good, but they were initially floored by the conversation of the Lancashire miners. 'The accent and the idiom were very hard for us, but we learnt quickly,' Weiss says. As, because they needed to, did Bevin Boys everywhere. 'Spoken at speed, broad Yorkshire was unintelligible, but it became part of you,' observes David Reekie. 'Phrases from the time still come into my mind: "Addle thy brass?" – how much you've been paid?; "Tha man san it up, lad" – pick it up.' Phil Yates in Pontefract became so adept in the dialect 'that when I went back to the solicitor's office in Winchester they couldn't understand me!' 'It was surprising how you absorbed the dialect,' says Doug Ayres.

In the pit some haulage staff would chalk phrases on wagons such as 'wonanatannerarowyerdunnera' – one and sixpence I owe you, don't I? – or 'getferknowthecaint' – find out how many wagons-full have been produced, and you didn't have to stop to translate it. In the pit, dialect wasn't something you put on when you met your mates, you said 'What see surree' quite naturally, which isn't so far from the Shakespearean 'What say Sirree' or 'dusteer' for 'listen'. I liked the expression that followed flatulence: 'Ta arse on fire?' Shakespearean again.

Ayres and another Bevin Boy sometimes amused themselves when they left the colliery and caught the bus into town,

by conversing in heavy potsy in the presence of rather posh matrons who didn't approve of us common workingmen, and then suddenly switching over into our best southern accents: 'I say, Ivor, are you going to the Victoria Hall in Hanley next week? Barbirolli is coming to conduct the Hallé Orchestra with Eileen Joyce as soloist.' And the grand ladies would realise they were being sent up and purse their lips.

If the miners' ways of talking seemed odd to many Bevin Boys, so did the way many Bevin Boys talked seem to the miners. Asked where he came from Stan Payne replied: '"London." "London. Ah thout thee war. Are Cockney then?" "Yes." "Ah. Thout thee war, joodging be t'twang." Me, with a twang? Surely only countrymen and folks north of Watford spoke with a twang.'

The speech of educated Bevin Boys particularly from the South wasn't always comprehensible to the miners: George Poston at Handworth Nunnery near Sheffield 'was once sent to phone the under-manager because the belt had stopped and couldn't be restarted. After a brief conversation he told me to get off the phone and get someone who could speak "boogering English".' And a matter of good-humoured mockery. When Doug Ayres spoke normally he was invariably asked: 'Why don't they talk proper?' John Squibb remembers a Bevin Boy in training saying in loud upper-class tones as they assembled for their work clothes: '"I have come for my togs" – the miners found that hilarious'. As they did Owen Jones's extreme politeness – soon after he arrived he failed to undo a haulage clip, said 'I'm awfully sorry, it was rather tight', and became known throughout the pit for it (as he did for his insistence on calling his snap tin his lunchbox). But miners were sometimes intrigued by the unfamiliar, as was Big Joe the morning that Geoff Baker arrived at the coalface and said cheerfully, 'I say, chaps, frightfully cold this morning.'

For days he practised saying it until he got it off and I said, 'Time to go public, Joe.' And he did. 'I say, chaps, frightfully cold this morning,' he told the other men on the face. They looked at him in amazement. 'Bollocks,' they all replied in unison. That was the last time Big Joe said it.

It was a profound shock to less worldly Bevin Boys how much of the language underground was, as Arthur Gilbert puts it, 'the less-used language of the old Anglo-Saxon dictionary'. 'Three or four swear words to a sentence, everybody from the young lads up. Even the manager on his occasional visits was no exception,' says Geoffrey Mockford. 'Down the pit, the string of obscenities had to be heard to be believed, but it was understandable,' comments Peter Archer.

Miners worked in appalling conditions. Swearing was how they relieved their frustrations and remained cheerful. This was a period in which you hardly heard any swearing in normal life. It was interesting that out of the pit, miners observed the strict social code that applied across society that men didn't swear in front of women. However, down the pit there were no women!

A code that was strictly observed in the pit was that a man didn't use the word 'bastard' about another. Denys Owen remembers that the welfare officer who met him and two other Bevin Boys when they first arrived at Gresford made a point of telling them. 'First, he said, we would have to get used to the bad language, but were at liberty to return this in kind. Second, we were never to use the word bastard to a miner. Asked why, his reply was "Because most of them are!"' The calumny was, nonetheless, good advice: in the coalfields, while 'bastard' was liberally used as a general descriptive, if directed at an individual it was always taken in its literal sense – as Doug Ayres remembers.

At the end of one afternoon shift they were still drawing wagons away full of coal to get as much production as possible. This meant it was too dangerous to go up the sloping tunnel, and it was a 20-minute climb to pit bottom after that. This was ten o'clock at night and the men were impatient; you had to be quick getting out to catch the bus. The overman, Harvey, who had quite an intimidating presence, was holding the men back, they were muttering, and he said something like 'Silly bastard'. A miner called Arthur launched himself at Harvey with his pick in his hand with the intention I'm sure of killing him. The others grabbed him and held him back.

Many Bevin Boys refused to be drawn into emulating the miners' bad language. 'They tried to get me to swear, but I wouldn't,' says Bert McBain-Lee. 'The under-manager was a churchman, never said damn, and nor would I. They used to joke to him: "He must be one of your lot."' A staunch Methodist, Alan Munford 'hated the swearing. I was friends with several Salvationists and we stuck together. I don't think we were over-righteous – we just didn't swear.' But another Methodist, Jim Bates, did, eventually, and found it cathartic.

> As a good Methodist I wasn't going to indulge in bad language, but one day I must have been annoyed with something and shouted out, 'Oh bother!' An old miner heard me, came up and said, 'Ave a good cuss and purge thysen, Jimmy.' How expressive that word 'purging' could be, both intransitively, through getting it off one's chest, and transitively, in saying what you thought of someone. Henceforth when the wagons ran away, I purged.

Perhaps the majority of Bevin Boys purged, some more than others. George Poston learnt to swear 'like the proverbial trooper'. Victor Simons learnt to swear 'politely – a collier said he'd never heard anyone swear so politely'. And Kevin, a Bevin Boy at Oakdale, learnt to use the worst of all words – without a clue as to its meaning. Says Bob Kinnear, a Londoner who before call-up worked for the Royal Arsenal Co-op Society:

> Kevin came from Cheltenham and was very well educated. It was said that he was very musical and could make a church organ sit up and beg. No one ever heard Kevin use a swear word. One day Kevin's butty said to him, 'I don't know, Kevin, but this is a bastard place to work in.' To which Kevin replied in all innocence, 'Yes, Ned, this is a bastard place.' At the end of the shift with half the men at pit bottom, Ned and Kevin came out of the turn road and Ned asked him, 'What sort of place is this?' 'This is a bastard place,' said Kevin. And everyone knew the unbelievable. Kevin had sworn.

Looking back, John Squibb is amused by memories of the language with which he was once familiar. 'I'd never heard anything like it,' he says. 'I couldn't understand most of it at first. By the time I figured it out I just accepted it. Given the dialect and all, it was almost poetic!'

With time, communication problems resolved themselves and miners and Bevin Boys learnt about each other, principally when they got chatting down the pit on snap. Many Bevin Boys, like David Reekie, remember the main topic of conversation being politics; perhaps more that politics were rarely mentioned. 'I don't think the miners had real political views, just a legacy of inherited ill feeling, cynicism and great bitterness,' says Meir Weiss.

The dogs – and the horses – were certainly major subjects of discussion. And football. And sex. 'Football and the other one, they were the only topics I remember,' says Tony Brown. 'I knew nothing about the first and barely qualified for L plates on the second.' 'Football and sex,' agrees Harold Gibson. 'The second, with my upbringing, I didn't know a lot about. One morning I arrived at the station and they were talking about a miner who'd been charged with rape. I tried to look knowledgeable but I had to wait until I got home and looked it up in the dictionary.'

'We were teased about our girlfriends, what progress we'd made in seduction,' says Doug Ayres. 'I won't quote that dialect. This was met with hot denials and we'd snigger as adolescents do about sex, hiding our ignorance. And the older men would look at us and say, "What's they on abait? Ar't they on abait that theer, that as dogs fight o'er, atna?"' Meir Weiss and the other Jewish Bevin Boys from the kibbutz took more ribbing than anyone. 'A collective of youngsters of both sexes living under the same roof: the miners envisaged nightly orgies, free love. They kept on telling us that when passing our house they saw the windows steamed up from the heat generated by sexual exuberance.'

Whether miners chose to raise politics in conversation or otherwise, almost to a man they were Socialist in spirit, with a historic loathing of the Conservative party, which had been in power between the two world wars when they'd undergone persistent hardship – the ugly years of lock-outs, hunger marches and out-of-work miners singing to theatre queues. 'They talked of the Confuckingservatives,' says Denys Owen, 'from which you gather that the blue party was not too popular.' Geoff Baker was always 'the bludy Tory, though only in a bantering way'. Once it became known down the Empire pit in Wales that Deryck Selby came from Winston Churchill's constituency, he was henceforth 'Lord Epping: and I wasn't even a Conservative at the time'.[3] When the miner in whose home he lodged in Sunderland discovered that Peter Allen 'was a supporter of Churchill, he went out of his way to pick arguments with me. He accused Churchill of telling the miners to go back to their holes in the 1926 strike.'

Churchill, whose 'lion's roar' had galvanised the British people, was not the hero to the miners that he was to almost everyone else. They had long memories, especially in Wales where in 1910, as Home Secretary, he'd sent troops into Tonypandy to support the police after miners on strike over imposed piecework rates rioted when the owners attempted to bring in black labour; and they remembered his part sixteen years later in crushing the General Strike. 'Mention of Churchill's name in Wales was like a red rag to a bull,' says Les Wilson. 'The mention of his name was tantamount to blasphemy.' On both occasions, and in 1921 between them, when coal prices and profits slumped in periods of economic depression, the colliery owners forced the miners back to work on substantially lower wages, and, in 1926 (the year in which the overwhelming majority of conscript Bevin Boys were born), longer working hours.

Jim Bates never forgot a miner saying to him, '"Jimmy, you know what a miner is? A bit of cotton waste. Wipe your hands on it when you want it, chuck it away when you don't."'

There was a great deal else relating to the inter-war years of which Bevin Boys were apprised. Of times when men got only two or three shifts a week on which to keep their families and sometimes, having already started work below, were sent home, without earning a penny. Of pits, when there was full employment, that hired men for only three days and other men for the other three, forcing them all on to the dole for the days they didn't work – a method by which colliery owners dominated their workforce; of pits where as many men were laid off on a Friday as were taken on; of pits that blacklisted trade union activists, forcing them out of the area to other coalfields. There were dark tales of greed and exploitation, of dangerous skimping on safety precautions, of overmen being sacked for properly withdrawing men in the presence of afterdamp, of instructions given when a man was killed in a pit explosion for matches to be put beside him so that any inquiry would infer he was to blame, not the colliery. Allegations that colliery owners were directing work into shaly seams that wouldn't have been profitable before the war and saving the rich seams for exploitation after it – allegations that the government had investigated and found to be untrue, but which still persisted.

It was difficult for Bevin Boys to know how much of what they heard was exaggerated, or to form an opinion about what kind of men mine owners were; historically many had displayed a considerable degree of social responsibility, providing housing and medical care, schools and places of worship. Some Bevin Boys saw their pit owner on occasion, others not at all. It was a different story at Lepton Edge near Huddersfield: George Elliott was constantly about his mine. 'He was dead eccentric as well as mean, but the men were amused by him more than they were resentful,' says Geoff Darby.

The only time I remember them really falling out with him was when the roof was coming down and we needed more timber but he said lots of props had gone down and he wasn't going to pay for more. We went on strike for a day and he gave in.

He was too mean to stop them winding coal when he wanted to go down the pit, so he'd get the banksman to wedge the safety gates, jump on top of the cage and descent at tub-drop speed. Occasionally he slept on straw in a surface pump room. Come 6am and he'd be using his stick to get us down the pit as early as possible and not have men hanging about having a smoke, waiting for the last 'man-riding' cages. You had to watch the backs of your legs!

Whatever manner of men mine owners were, the mining communities had endured untold deprivations and many Bevin Boys, like Stuart Chislett at Bullcroft Main in Yorkshire, were shocked; and they came to understand why miners distrusted the government almost as a matter of principle, the wartime coalition included: politicians, they thought, always sided with the bosses. 'I'd been vaguely aware of some of the coal industry's problems but had no real idea,' Chislett says. 'As we gradually heard more my sympathies were entirely with the miners.'

Les Wilson, down the Llay pit in north Wales, had his eyes opened too. 'My Conservative views,' he comments, 'were slowly eroded like my lungs probably were.' He left the pit a confirmed Labour man and had trouble with his Tory-supporting family. '"I won't have you spouting that communistic claptrap in my house," my old mum told me.'

Bevin Boys also came to understand how the past explained why so many miners seemed to distance themselves from the war. 'They were patriotic, but they were committed to the war effort only in a general sense,' says Meir Weiss.

> What was written and broadcast specifically about it was pretty meaningless to them. They lived in their own isolated world. Many were not even interested in the war itself. When it was ending, it was very exciting, this town being taken, that town. The miners always took the newspaper down the pit, but they only looked at the headlines. They were only interested in the racing results. What were the greyhounds doing? That Paris had been taken was a mere detail.

*   *   *

The Porter pay agreement that had increased miners' pay in early 1944 had also raised Bevin Boys' wages to £3 10s at eighteen and took them to the miners' basic of £5 at twenty-one.

Miners' unions in some areas, however, had negotiated better deals; Bevin Boys who went to Nottinghamshire, for instance, were delighted to find the twenty-one rate was £5 18s – a rate from which those in the two Derbyshire pits of Bolsover and Creswell also benefited. 'The Bolsover colliery company owned four pits in Nottinghamshire,'[4] explains Victor Simons. 'Negotiating pay rates with two different unions could have been a problem. The company made matters simple by recognising the Nottinghamshire one.'

While the overwhelming majority of Bevin Boys worked on haulage and were employed by the collieries on the flat rate for their age, those who worked on or near the coalface as putters, fillers or (in Wales) colliers' assistants had their wages paid on a piecework scale by the collier they worked for, except for 5s that came from the government under the Porter award. Colliers were virtually all contract men, not colliery employees, as were rippers, packers and the men who drove new roadways, who also paid those working for them under the same system. The wages of these men were governed by a confusing number of factors. Whether the agreement with management was by the ton or the cubic yard of coal, for example, or just the shifts worked. Whether payments were made for the number of props erected, or the amount of 'muck' (rock) taken out, or the yards of conveyor belt dismantled and reassembled. Whether there was compensation for mechanical breakdowns when no coal was moved. The price for the same task could vary from one coalfield to another, or from seam to seam in the same pit, or even from district to district on the same seam.

Like all other miners Bevin Boys were entitled to extra payment for working in water ('wet money') but, as with everything else, the rates were negotiated locally. In the Louisa pit in County Durham, John Cook got 7s 6d a week; in the Handsworth Nunnery in Yorkshire, George Poston got three pence a day 'boot money' for working in 6in of water – a weekly 1s 6d. 'We also had water dripping off the roof, for which we were supposed to receive sixpence a day "clothes money", because we got wet both ways. But the management took the view that one compensation was sufficient and chose the cheaper option.'

Colliers traditionally tipped those working for them, but what a Bevin Boy got depended both on what the individual miner earned and his generosity. Two to five shillings was about average, but Reg Taylor remembers a collier called Fred Larthorpe who was more open-handed,

> especially if he felt you'd been putting your back into it for him and it was coming up a bank holiday. Then he'd give you 15s or even a pound, which was incredibly generous. To earn his money and mine meant filling eight 5cwt tubs a day: two tons. It took some doing. There were a few exceptional coal-getters [colliers] who did more but most, including Fred, couldn't have earned much more than £6.

'Some men gave you a lot more than others,' says Geoff Darby. 'What they gave you depended, it seemed to me, on whether they liked too much to drink.' Which was why some wives showed up at the colliery on payday. Dave Moody remembers a lot doing so 'and they took the pay packets straight off them, otherwise they'd be in the pub just down the road'.

Most collieries had moved to paying out in sealed brown envelopes (in the Lady Victoria in Scotland, George Ralston remembers, wages were dispensed in small round tins, to save paper). All the flat-rate Bevin Boys collected theirs from the window of the pay office, as did many of those on piecework. But some collieries still stuck to the old ways, all the money paid to the miner, who then handed over his Bevin Boy's portion. 'Waiting for your one allocated face miner to pay out was okay,' says Geoff Darby, 'but there were times when you worked for more than one in a week, even a different one every day, so waiting felt pretty slavish.'

Miners frequently disputed what they were paid. 'There'd be huddles of miners around the pay office area studying their pay slips and counting their money, always grumbling and arguing over the contents,' says Norman Brickell. Sometimes there was more than grumbling. Geoff Rosling: 'The arguments could be extremely violent verbally, but never reached physical violence. The overman, being a company man, obviously beat down the colliers' claims and they'd end up eyeball to eyeball – but that evening would be drinking together in the miners' institute.' It was only in the institute that the collier whom Dennis Faulkner assisted would give him his pay. 'I had to go there on a Sunday morning, otherwise I didn't get it. The institute was the only place open on Sundays that served alcohol. If he was in a good mood he'd buy me a small cider, sometimes.'

Bevin Boys who became colliers themselves could earn good money: the best week Alan Gregory ever had his gross pay was £11 11s – this at a time when an average job in civvy street paid around £4 10s. Others who worked with rippers and road drivers – men who could earn more than colliers, as much as £20 a week – often did well. The ripper Jim Ribbans helped regarded him so highly he split the money 60:40.

Stoppages were a bone of contention with most Bevin Boys, particularly those on minimum wages who could least afford them. Derek Agnew listed what was taken out of his basic haulage hand's money at Chislet in Kent. He accepted the necessity for the doctor (6d) and hospitals (4d) in case of accident or illness, was happy enough about the 2d for the Red Cross, though he would have liked to have been asked if he agreed, and thought the 3d for welfare was reasonable since it covered use of the recreation rooms ('though we always pay extra for a film show or dance'). But he was annoyed by the 8d taken for the pithead baths (though acknowledging most of this was for soap) and was 'completely flabbergasted' by the 6½d charge to cover the carbide he needed for his lamp down his naked-flame pit. 'So we not only pay to wash off Chislet's coal but to light up the pit, too,' he wrote indignantly.[5] Altogether, including PAYE and health insurance, Agnew was docked 14s 4½d, leaving him with £2 11s 4d.

What deductions were made for, and for how much, were decisions taken at local level. Typically there was a contribution towards the cost of the canteen if there was one (6d), various benevolent funds (2d), aged miners (2d), the miners' club or institute (1d), and the colliery band to play at an individual's funeral should the need arise (1d). At the majority of collieries a deduction was made to pay the assistant to the checkweigh man put up by the union; and while at many collieries miners paid the pick sharpener themselves for whatever work he did for them, at many more everyone paid an obligatory levy instead to cover the cost. Both stoppages annoyed Bevin Boys on haulage who gained nothing from the employment of the one and had no need for the services of the other.

But nothing annoyed many as much as a stoppage about which, oddly, Agnew made no mention: the cost of belonging to the miners' union. Many Bevin Boys objected, not so much because of the sixpence or as much as a shilling that was taken from them weekly, but because they had tried to refuse to join. Some did so as a matter of conviction; others because they resented being coerced; others, many of them without a view one way or the other, because they worried that union membership would in some way lock them into coalmining and make it even more difficult to get out. And there were those like Geoffrey Mockford and his colleagues sent to Pleasley colliery in Derbyshire who hoped that if they weren't in the union the miners would refuse to work with them and they'd be on their way to the forces. It didn't, of course, work out like that. 'I was asked by letter to join the union and said no,' Mockford explains. 'Next week 10d was deducted from my pay anyway. I complained but it made no difference, it was still taken off – and continued right up to the end.'

According to an assurance Ernest Bevin gave to the House in early January 1944, there was no compulsion on his part, nor was there any statutory obligation, that a coalminer should be a member of a trade union. What he said was true as far as it went. The reality was that in many parts of the country union branches operated a virtually closed shop. In late January the Conservative MP for Glasgow Pollok, Thomas Galbraith, pressed Bevin on the matter, saying he understood that in south Wales, union membership was compulsory.

Bevin was in a spot. There'd been difficulties enough getting the Bevin Boy scheme up and running and the first batch of conscripts were still in training. Had he agreed that what Galbraith said was true, voices of protest would have been raised on all sides; had he said it was true but that Bevin Boys should have the right to refuse, there was every possibility that the miners in south Wales would take strike action. In the circumstances, Bevin obfuscated, as politicians do. He simply quoted from the agreement with the South Wales Miners' Federation, which stated that 'every workman normally employed' in the coalfield was required to be a union member – and left it open to the interpretation that as Bevin Boys weren't *normally* so employed, they didn't have to be. It wasn't until 1947 that George Isaacs, the then Minister of Labour and National Service,[6] was able to be more straightforward, admitting in the House that in 'some few collieries' union membership was a condition of employment.

Bevin Boys served in about 350 collieries, rather more than a quarter of those in the country.[7] How many 'some few' were is impossible to answer. The Welsh miners' union had a blanket policy;[8] other regions were less hard-nosed and left the issue to branch memberships. Some of these were uncompromising and Bevin Boys were pressured into joining (arriving at Kibblesworth in County Durham Bill Hitch was told '"Go down the hut where they play billiards, see the union bloke and sign up" – it was an order, not a discussion'); others enrolled new arrivals automatically but allowed them to opt out if they chose; others were relaxed about the whole business. Roger Spencer at Tursdale colliery in County Durham 'was never in the union and it was never suggested I should be. The miners knew we were only there because we had to be.' As they did at Welbeck in Yorkshire. Stan Payne:

I'd like to record that we were put under no obligation to join the union. The union membership had decided, as we were there by government direction, we should have free choice in the matter. And we were free to enjoy all the benefits of membership – the use of the club's facilities, subsidised beer and cigarettes – and not contribute a thing. I believe without exception all the lads joined, and there were 40 or 50 of us spread around the village.

At Lepton Edge, also in Yorkshire, Geoff Darby remembers the miners taking the same considerate stance but the Bevin Boys' attitude being entirely different:

> There were 52 of us among the 300 miners and only two of us joined the union. The others didn't want to be identified with the mining industry. Some of them just refused to recognise the miners. One chap from the same grammar school as me had a chip on his shoulder like a pit prop. I joined because I believed in unions; my father was always a strong union man at the Post Office.

Most Bevin Boys with a family tradition of union membership expected to be union men; others were happy to be included because they considered it right – Tibby, the miner-fitter that Ian McInnes worked with at Snowdown in Kent, told him: '"If you enjoy the benefits negotiated on your behalf you should pay your dues", and that seemed a good argument to me.' Some Bevin Boys went beyond simply accepting their membership: they threw themselves into union affairs. Desmond Edwards was elected to the Tower production committee in south Wales, Geoff Darby to his union branch committee, and he walked the 5 miles to and from his home in Huddersfield to the pub near the colliery where meetings were held. And 'somehow I got the job of writing letters for those miners who couldn't write or at least had difficulty reading.'

> I sorted out papers about their entitlements to extra rations – all sorts. I eventually got to the stage where they were coming to me with personal problems, telling me about marital difficulties, asking me to talk things through. I used to think: I'm 18 years old, I know nothing much about all this. But I got involved because I cared. I vividly recall the times I went by bus to Wakefield to see a miner called Joe who was in the miners' hospital after an accident in the pit. He was killed later. Only a youngish chap.

Practically speaking, it made sense to be in the union: a man who was and sustained injury or was too ill to work received financial support; a man who wasn't had only the 3s 4d a day paid by the government, to a maximum of £1 a week. When he was crushed between tubs in the Louisa pit in County Durham, John Cook regretted not bothering to join: he was off three weeks and had to borrow money to pay the hostel. When Jim Ribbans transferred from Wales to Snowdown and found he didn't have to be in the union, he didn't bother either – but when he badly injured his hand was forced to return to work before it was better because he was desperate for money. Geoff Darby blessed his lucky stars that he was a member when a tub of rock overturned and damaged his foot. First, he got a payment from the union; then, when the Mineworkers' Indemnity Association doctor twice said he was fit to work and he wasn't, the union paid for a specialist to examine him, and he was awarded compensation. 'What a revolting little squirt the association doctor was,' Darby says.

I had my accident on August 7 1945, the day after the VE Day holiday. When I got out of Huddersfield Infirmary I had to travel to Dewsbury to see him. He returned me to work on October 15, though any fool could see I wasn't right. My diary shows I was back in hospital on November 10. On December 27 he returned me to work again – and on January 8 I was back in hospital for another operation.

Stan Payne and his fellow Bevin Boys at Welbeck were grateful for the union's protection on the day they got their first week's wages:

We were paid the Ministry of Labour rate, which was lower than the going local rate. A visit to the cinema in Mansfield or Warsop, *or* a pint, *or* a packet of Woodbines meant almost total depletion of funds. Our protest fell on stopped-up ears as far as the pay clerk was concerned. He interpreted the regulations to mean that, as we were drafted, our pay rate was fixed by the MOL.

Here steps in one Henry Foster, secretary of the Welbeck branch. Ten-minute conference with management. Matter resolved – we got the rate for the job. Not a fortune but surely justice. What we never got afterwards was a friendly smile from behind the pay window.

Those Bevin Boys who chose to ignore the union perhaps only appreciated its influence when it came to their wages. George Poston transferred from Yorkshire to Ellistown colliery in his home county of Leicestershire, wasn't obliged to sign up with the union there and didn't – and got a shock on his twenty-first birthday when he was expecting the miners' full pay entitlement: the pay office told him he wasn't getting it unless he became a union man. 'This seemed like blackmail to me, but there was little I could do about it. The joining fee was ten shillings, which could be paid at sixpence a week, so I paid across my first sixpence, entitling me to a union card.' George Ralston hit the same impasse at the Lady Victoria in Scotland. Indignant, he sought a meeting with the colliery manager who told him he was sorry 'but he couldn't help me, it was a union decision'. Ralston wasn't prepared to leave it there and wrote to the Federation headquarters in Edinburgh. To his surprise an official was sent to see him, but the position remained unchanged, other than that the Federation was prepared to waive the joining fee. Like Poston, Ralston gave in.

In his 1947 'some few' admission in the House, George Isaacs revealed there had been Bevin Boys (he gave no figure) who'd held out against union membership in collieries where it was rigidly enforced and they had had to be transferred 'to other collieries where this condition of employment did not apply'. It's unlikely there were many who felt so strongly.

Unquestionably the worst clash between Bevin Boys and union occurred at Bullscroft Main in south Yorkshire, where feelings ran so high that every Bevin Boy in the pit walked out – and never returned.

Like many other union branches, Bullcroft's hadn't take a rigid line on membership and for eighteen months, Stuart Chislett says, 'we all rubbed along pretty well'. Then the secretary of the branch sent the Bevin Boys a letter:

Dear Friend,

I think by now you have almost become accustomed to your new sphere of working life (not a very kind way of getting a living, is it?)

Anyhow, I am instructed to inform you that everybody working underground at this pit are in the Union with the exception of a few of you newcomers to the industry, and we think that you have now been here long enough to have joined the Y.M.A.

So don't be different to all the rest of the pit, and please come and join at once, otherwise unfriendly decision may be taken against you.

Signed on behalf of the committee

Perhaps the committee meant only something minor, such as the withdrawal of the privilege of using the club, for example. To the Bevin Boys the intemperate language signalled a threat. 'Revolutionary' meetings quickly followed. At first the Bevin Boys withdrew their labour. Then when management failed to find a way out with the union, many went picking swedes for a local farmer ('in the depths of winter, snowing, and bitterly bloody cold') to pay for their keep. Others departed for home and eventually so did everyone. Adds Chislett: 'I don't remember the course of events, but I do know that not one of us went back to Bullcroft. We were rounded up and posted off to the army – the official communication I got informed me that my transfer had been approved, which is one way of putting it.'

Having been a coalminer, Chislett 'was a natural for the Sappers' and spent the next 2½ years on bomb disposal, dealing with UXBs and clearing minefields that had been laid on the beaches in 1940. 'You might think it was a case of out of the frying pan into the fire,' he says. 'As far as I was concerned it was "Good old army!"'

Seven

Life's What You Make It

The Ministry of Labour and National Service sent Bevin Boys from coalmining areas for training locally. Bevin Boys who came from non-mining parts of the country finished up anywhere. After training, all Bevin Boys were theoretically able to choose a pit, but without a guarantee of going there. Most of those on their home patch wanted to work locally, which made sense from everybody's point of view: they could live at home and it took pressure off scarce accommodation. In choosing a pit, it helped to be in the know – which Jim Ribbans, like other outsiders in south Wales, wasn't when he finished training at Oakdale. 'I couldn't tell one pit from another and some of the names were incomprehensible. In the end I picked Werfa Dare at random. At least I could pronounce that.' Welshman Desmond Edwards was in a different position. He lived within sight and sound of the Nant Melyn pit at Bwllfa in south Wales, but elected not to go there, instead choosing the Tower at Hirawun, not just because his father worked there but because the Nant Melyn, although it had modernised and introduced coal-cutting machinery, 'had a reputation as a death trap – while the old-fashioned Tower, where coal was hand got, had a nine-foot seam a man could stand up in and an excellent safety record'.

A fair proportion of those from outside mining areas got posted where they wanted. Alan Gregory went to Cossall in Nottinghamshire because it was reasonably close to London (and he'd found out the pay there was better than average); John Wiffen went to Gresford in north Wales because it was a not-too-distant bus ride from his home in Cheshire; Peter Archer went to the Old Coppice in Staffordshire: handy for home in Wednesbury but, more importantly, 'my cousin was there as a volunteer – and he had a car'. Perhaps David Reekie was unique in knowing where he was going even before he set off for training – he was very vaguely connected by marriage to the owner of the Horse Riggs colliery in Yorkshire. 'When my brother-in-law discovered I was going to the mines, he asked me if I wanted to go to Yorkshire,' says Reekie.

'If you're a fair example of Yorkshiremen, the answer's yes,' I said. So he contacted his uncle, John William Buckley, who was a colliery owner only in a very modest way and someone I'd met once or twice as a schoolboy. He agreed to take me on. I saw him at his large double-fronted stone house in Hipperholme. 'Don't expect any favours from me, lad,' he told me. 'Just do as you're told and do your job right.' And that was that. I only ever saw him at the pit once. Anyway I was grateful to him and more grateful to the Ministry. I thought they showed some heart in the matter.

Denys Owen was another who thought the Ministry showed heart, though he didn't get the pit he thought had been fixed up, as in Reekie's case by personal contact. The manager of Hafod colliery near Wrexham was a friend of his father's, who'd said he was desperate for men and would welcome Owen with open arms. When it came to it, Owen found himself on the way to Gresford. He didn't mind: the two school friends who were supposed to be going to Hafod with him had also been switched – and the Ministry had kept their side of an unusual bargain. 'It's an odd tale,' says Owen.

The three of us had birthdays close together and soon after we found out we were destined for the mines we all received a letter asking us whether in view of the serious manpower situation and the shortage of fuel we'd agree to go earlier. You can imagine our reaction. But wiser counsel prevailed in the person of my father. He'd been called up in the First World War but in view of his age was now working in the Civil Service. He'd become accustomed to the Civil Service mind and he suggested we agree provided they kept us together. And despite the change of colliery, they did.

Not all collieries had manpower shortages and the hopes of many Bevin Boys were dashed, including most of those from the South East who had their eye on the Kent coalfield – which, with only four pits, took by far the fewest men.[1] The unflagging Mass Observation researcher Eric Gulliver recorded the case of a fellow trainee at Swinton near Manchester who was getting married in a month and who in anticipation of at least staying in Lancashire had already put a deposit on a house, and was devastated to be posted to Staffordshire. After training at Askern Main near Doncaster, Bill Hitch asked to go to nearby Bullscroft because his landlord was a miner who worked there and he could continue living where he was, but was sent to Kibblesworth in County Durham. He thought the system perverse. 'The Scotch lads were sent to Kent and the South lads were sent North and Scotland. It was a farce, asking.'

Under the Essential Work Order, only miners and those employed in shipbuilding were allowed to move between workplaces and then only within their industry. Initially Bevin Boys were supposed to stay at the colliery to which they were assigned, but on occasion regional National Service Officers responsible for the order's implementation were flexible, if a man had found another colliery to take him. After the war ended and restrictions on the movement of labour were scrapped, a good proportion of Bevin Boys changed pits.

Most who did wanted to live at home or close enough to get there at weekends. It wasn't always a smart move, as Jack Garland discovered when after two years at Markham colliery in south Wales he organised a transfer to Bromley colliery near his home in Bristol. 'In Wales I actually got a sense of satisfaction from my work. It wasn't the same in Somerset,' he says.

> In Wales I worked with one man on the coalface the whole time. We cut the coal and we loaded it. I was strong, I could hew with the best of them. Me and my butty were a team. In Markham I was moved around doing different jobs and there wasn't any sense of pride in it. And while it didn't have gas, Markham was a soaking wet pit, low and very hard work – you had to tumble the empty tubs off the rails to get the full ones passed and then tumble the empty ones back on. One day at the pit bottom I noticed a lump in my groin. I'd given myself a hernia.

Ronald Griffin moved pit less because he wanted to be at home but because he was so unhappy. The biggest pits absorbed a couple of hundred Bevin Boys; the majority had anything from a dozen to eighty. A few, however, had only a handful and fewer still just one; and at Aberbaiden, in the village of Aberkenfig near Porthcawl, Griffin was it.

He made a bit of an entrance to the village: the local milkman in cow-gown and breeches offered him a lift from the station and he came in standing behind the milk churns on the dray. That, such as it was, was the high spot of his time in south Wales. His relationship down the pit with the doggie [foreman] didn't improve after he refused to buy a shovel; it seemed to him 'that no one ever used my Christian name, it was always "Him" or "Boyo" or "Mun" – or mostly nothing at all'; and he thought the stale air underground was burning his lungs. He was well looked after at his lodgings, but he was very lonely.

> There was nothing to do in the evenings but watch streams of army convoys along the road to Port Talbot and Swansea – this was June 1944, the build-up to D-Day, though no one was aware of what was going on. Porthcawl was a seaside town with a mined beach and nothing to offer but the sea. Sundays in Welsh Wales were a death sentence: pubs and cinemas closed, even tearooms. Nothing but chapel and rain.

Griffin wrote to several Midland pits to see if they'd have him. Baggeridge near Wolverhampton agreed, provided he got the requisite permission.

When Griffin saw his regional official he got a roasting for seeking a transfer after only three months but was let go nevertheless. Looking back at the combination of circumstances that made him want to get away he says: 'I suppose it was largely in my mind, but things had built up to the point where, if I'd had to stay, I think I would have become seriously ill.'

*   *   *

Few of the first Bevin Boys who in early 1944 went off to dig for victory in a way they could never have anticipated had hostels to go to. Even when all forty-four were eventually provided there were never enough places for everyone.[2] In Wales there were always more men in lodgings than in hostels.

Managed by the National Service Hostels Corporation on behalf of the Ministry of Fuel and Power, hostels were basic, but no more so than the average army camp from which they were either adapted or on which they were modelled. Some offered more facilities than others (the biggest ran to a hall with a stage and doubled as a cinema), but all were similar in provision and layout, consisting of a welfare block, dining room, tea bar, lounge, games room, quiet or reading room, and a sick bay staffed by a nurse and in many cases a matron. The living blocks of Nissen huts, with their familiar green-painted semi-circular roofs, each slept twelve men and were linked by walkways to clusters of toilets, baths and showers. A 4in pipe just below the curve of the ceiling provided each hut's heating: inadequate in winter but useful for drying clothes. Unlike in the army, residents had domestic staff to make the beds and do their laundry if required (an extra 2s or half-a-crown), and, unlike in the army, they had no parades or kit inspections. Those going on the day shift got an individual early morning call too from the 'wakey-wakey' man: the night watchman came around to give them a shake and shine a torch in their eyes.

The weekly charge for board and keep was 25s (up to 30 in late 1946), with men who went home at weekends entitled to a 5s reduction. On payment, a resident received a weekly booklet of meal tickets that gave those in full-time residence fifteen main meals, two on weekdays and three on Sunday. Hostel food was plain, but it was plentiful: a surprise to Bevin Boys used to wartime rationing. Breakfast was available from midnight to around 10a.m. to cater for the various shifts worked at different collieries. The main meal (early evening for the majority) was available throughout the day. The tea bar, from which snap could be ordered, had long opening hours and the afternoon high tea available from it (5d) was reckoned to be the best value on offer. For a Bevin Boy of eighteen on a wage of £3, these outgoings in a six-day week could come to 37s 6d, on top of which, for most, there were bus fares of a shilling or one and six. With stoppages of around 15s, that left 6s spending money, not taking account of any essentials. Wage increased at nineteen, twenty and twenty-one but, while there were exceptions, most Bevin Boys never reckoned on having much more than 15s to call their own.

A Bevin Boy able to live in a hostel was sure of what he'd get for his money. Lodgings were an altogether chancier business – if indeed, in many places, they were possible to find. It wasn't easy for conscript Bevin Boys in particular: the first ex-servicemen who'd transferred to mining had had a sixteen-month head start, and even before they got there, in some areas, particularly the Midlands, tens of thousands of evacuees had beaten them to it.

Ex-RAF volunteer Eric Gulliver was too late to enjoy the comparative advantage of earlier service personnel: he arrived in Nottingham in July 1944, with another Bevin Boy, to work in the small Clifton colliery. It was 8.30 in the evening. The pair tried the Salvation Army for help, then the police station, then phoned the colliery. Then they tramped the streets of little back-to-backs in the Meadows and Gedling areas around the city's two collieries without any luck. A number of landladies were prepared to take them in – B&B only; one offered full board for £3 5s, which was 5s more than their wages would be. After nearly six hours of knocking on doors they gave up and returned to the police station. The police officer on the desk thought the situation was disgusting and advised them to go home to London, which they did. From King's Cross station they sent a joint letter to the colliery, registered post:

Dear Sir
We did not report for work this morning (20.6.44), our reason being that we were absolutely unable to obtain accommodation anywhere in Nottingham yesterday . . . Consequently we had no alternative but to go to our respective homes in London and trust that as soon as you find available accommodation you will let us know.

Gulliver didn't leave it there. It was now the early hours but he went to the *Daily Mirror*, where he told his tale to the industrial reporter, then to the Ministry of Fuel, which directed him to the mines division near Westminster. Here an official told him he was in breach of his reporting instructions and that in Nottingham he should have gone to the welfare office. Gulliver pointed out that it was shut. In which case, the official told him, he should have slept on a park bench. Next stop the Ministry of Labour and National Service. Another official, sympathetic this time, but unable to suggest what to do.

Five days later Gulliver and his fellow Bevin Boy received official word that their complaint had been dealt with and a request to return to Nottingham. The welfare officer met them off the train and escorted them to the colliery, where they were warmly welcomed. The accommodation that had been found for them Gulliver deemed 'excellent'.

The matter had repercussions. The *Daily Mirror* had been phoning around. On 6 July, Lewis Silkin (Labour, Peckham) asked Bevin in the Commons whether he was aware that men were being directed to work in the mines 'without accommodation on terms within their means being first arranged for them'. 'I am not aware of any general difficulty of this kind,' Bevin replied. 'The instructions are that, when men are sent to work away from home, arrangements shall be made to ensure that satisfactory accommodation is available to them.'

Quite a few Bevin Boys up and down the country might have liked a word or two on the matter with the Minister of Labour and National Service.

Local officials in the mining areas, trying to accommodate newly arrived Bevin Boys, were as up against it as those responsible in the training centres. Most collieries made an effort to keep a finger on the possibilities in their direct communities, providing a list of mining families or widows of miners who'd expressed themselves willing – but the list was often dishearteningly out of date, vacancies long since filled. Whatever way Bevin Boys got fixed up, the situation was painfully familiar to many from what they'd already experienced: houses that were badly run down, without electricity or indoor plumbing, and an outside privy, its door with a large gap top and bottom for the wind to whistle through. Invariably, too, lodgings were 5s more than the hostels, to cover daily snap and laundry, which in the hostels were paid for separately. While what was on offer might be anticipated, lodgings sometimes had shocks in store. Michael Edmonds was horrified on his first night to have to sleep 'not with my fathers but, to transfer a biblical expression, with my landlord'. David Reekie feared the same fate, or worse, on his first night in the home of a family of four who all worked in a woollen mill. The first to go to bed, he was directed up a flight of stone stairs at the head of which was a double bed. Could it be for only one person, he wondered, and stayed wide awake with the covers to his chin.

> The sound of the first set of clogs ascended the stone steps. One grown-up daughter appeared, crossed to a door behind which I saw was another flight of stairs. Then came the second and then the mother. The old man was last and I lay in dread. I could smell him before he hove into view. Fortunately he continued his upward path.

Reekie left before breakfast and, like Edmonds, didn't return.

Denys Owen's first night might have figured in a *Carry On* film. Come to be employed at Gresford pit near Wrexham, he found himself split from his school friends – the welfare officer had been able to find lodgings for only two of them together, and on his own he was dropped off, in the dark and in deep snow, in the hill village of Rhosllanerchrugog. Frozen and tired out, he eventually found the address, only to be told there were no vacancies, but that a woman at the other end of the village might have something. She did. Owen tried, and failed, to eat a supper of cow heel watched by the three women in the house, 'the landlady in her fifties, her daughter in her thirties and a thin girl of 19 or so who scratched herself endlessly'. He was sent to bed with a candle.

As I went up the stairs another Bevin Boy came down, going on night shift. 'Lock your bedroom door,' he said, though he didn't say why. Bizarrely, someone had sewed up my pyjama bottoms during supper – one of the women, presumably. Trying to put them on I tumbled over just as there was a ra-ta-ta-ing on the door, accompanied by shouts of 'Let us in.' Whether there were two or three of them I don't know: luckily I had locked the door. No mention of it was made again. I pleaded with my friends to find somewhere we could all be together. In the meantime the daughter, a very large lady whose husband was fighting in France, made it clear she had designs on my body. It took six weeks to find a place. A very long six weeks.

Many Bevin Boys had rueful experiences. More than a few either shared a double bed or shared a single by working different shifts; Peter Allen in Sunderland shared a double with another Bevin Boy squeezed into a room in which a lodger shared a second double with the son of the household. Others had experiences more dire: beds never made, sheets never changed, worse, fleas and bedbugs. And cockroaches. In his digs in Ynysybwl from which he travelled to the Lady Windsor colliery near Pontypridd, Alan Lane remembers 'the kitchen floor crawling with them. "Don't stand on them, they make a mess," the landlady told me, "they'll go back when it's daylight."' He got used to them (as he got used to the jumble of jam and cheese and dry bread she gave him for snap). He never got used to the little patch of damp that regularly appeared in his bed – 'the landlady's married daughter came round most days and they put the baby in it'.

Sent to South Hetton colliery in County Durham, Brian Folkes and another Bevin Boy were billeted 'in a decent council house which internally was a slum'. He adds:

I couldn't believe it. The husband worked as a collier at an adjoining pit and was on £9 a week, which was good money, the equivalent of maybe £500 today. But none of the upholstered furniture had any seats left; even the family bed had a big hole in the middle. And there were no beds for us – the Ministry of Labour supplied two truckles and a couple of army blankets each, beneath which I shivered. An SOS home brought an eiderdown to save me from permanent frost damage.

The house did at least have a fixed bath. The bathroom was off the kitchen, with a gravity hot water feed from the living room back boiler. The fall was negligible – it took 20 minutes to run a bath. The husband had pithead baths so he didn't use it and his wife wasn't much interested in hygiene. So it was ours. It was the only saving grace.

Many landladies were elderly and simply didn't know how to communicate with the young men who entered their homes. In Mansfield, Derbyshire, Geoffrey Mockford's landlady, who might have been in her eighties, 'kept herself to herself in another part of the house, and we had nothing to say to each other anyway. There was no newspaper, no radio. For all I knew the war might be over.' In digs outside Doncaster with two other Bevin Boys, Alan Munford hardly ever saw his landlady, the widow of a miner killed down the pit, in a house that had no clock on view. 'No one had a watch in those days and we never knew the time,' he says.

> Being a farm worker I was used to getting up early, so I'd get up at what I estimated was about four and walk to the railway level crossing to look at the clock in the keeper's box. When I got fed up with that, I went to Doncaster labour exchange and indented for an alarm. It took about a month before I got a permit to purchase one.

Some landladies had no idea how to feed their young charges properly, or were unable to have a breakfast ready at 4.30 or 5 in the morning, or couldn't cope with the rationing system, or were simply unscrupulous. Bevin Boys in hostels went to work on a breakfast of porridge and a choice of dried scrambled egg[3] on toast or Spam and dried egg fritters or bacon or sausage, and as much tea and toast as they wanted. Some Bevin Boys in digs got a slice of toast or perhaps two (or just bread and marge) and a cup of tea. Some got nothing at all. Peter Allen's stomach turned on the mornings his landlady put fried tripe in front of him. Many Bevin Boys never saw a drop of their national milk ration, only ever getting dried – and saw little or none of their cheese. 'We handed over our ration books with the extra coupons but none of it ever came our way,' says Alan Munford. 'It would have been nice to have a bit in our sandwiches, but for snap we had jam, jam and jam. We never asked. You don't think of things like that when you're 18.'

For some, main meals were meagre (one Bevin Boy is on record as saying he was always served salad, summer and winter, most of it potato) or almost inedible (many Southerners in County Durham never did come to terms with cold cooked tripe in vinegar, although it was a favourite with Tynesiders. John Squibb remembers 'the old terrier under the table doing very well'). For Geoffrey Mockford,

> a regular meal was a square of toast with three sardines on it. I don't remember a sweet but probably had something. One Sunday I remember being given a piece of Yorkshire pudding as a starter, the main course was the same pudding with a piece of meat, and for sweet the same pudding appeared yet again, but this time with some jam on it.

The fish and chip shop was the savour for many. Fish was never on ration but even for the diligent housewife was hard to find. It was, however, available at the chippie though, as there was a shortage of wrapping paper, it was often necessary to take your own if you wanted your purchase wrapped. If funds were low and a piece of fish was out of the question, in the North there was always the cheaper scallop – two thin slices of potato with a tiny morsel of fish between them, dipped in batter and deep fried, served, if desired, 'with the bits on' – what floated in the fryer. If funds were virtually non-existent then a few pennyworth of chips had to suffice. Geoffrey Mockford carefully spaced out his visits to the chippie – he had to catch a bus into Mansfield town centre, the fare bringing his usual expenditure to 8d which, small a sum as it seems, 'was a drain on my finances'. Doug Ayres remembers agonising in the War Workers' Club in Stoke whether he could afford to spend 'thruppence on a buttered bun which I really fancied'.

Initially as unhappy down the Pleasley pit as Ronald Griffin in Wales, 'working at the pit bottom with people with whom I had nothing in common, which made conversational pointless', for two months Mockford didn't even know another Bevin Boy. Getting back to his lodgings at 5p.m. and going to bed at 9, he felt 'a complete sense of isolation and began to doubt my ability to keep going'. Then in a chance conversation in the street he learnt that a hostel was about to open in the town – and he moved in with a huge sense of relief. 'Not only did I get decent food at last and plenty of it, I got something better – I was among my own kind. I had other people to talk to.'

Not all Bevin Boys in poor lodgings had the option of moving into a hostel. Some didn't want to anyway. There was no hostel in Stoke when Geoff Baker went to Staffordshire but even had there been (as there was later) he wouldn't have chosen to be resident:

> I'd been in boarding school for nine years and I didn't want to be part of some sort of boarding set-up, which the hostels were. I was a miner and I wanted to live the life of a miner, among miners. I wanted to be with the professionals, not the amateurs, which of course I was, though I did a reasonable job.

While he was in training, Baker only had use of a bed that was a night shift worker's by day, and 'thinking it was time I had a bed of my own' he kept on the lookout to move to other lodgings. One day a Bevin Boy from Birmingham who was transferring from Norton colliery to one nearer home told him 'what a damn good landlady he was leaving. Her name was Lucy and she had a lodging house in Waterloo road, Burslem. I nipped round smartish.' Baker thought she seemed quite old ('in her thirties, I imagine'), liked her immediately and took the vacancy. 'Lucy was a smashing person, with a sense of humour, which was just as well because I led her a ragged life,' he says.

A fair while after I was there she told me she was putting the lodging rate to 35 shillings. 'Sorry, Geoffrey,' she said, 'I can't make do on 30 bob.' Swine that I was I thought, I'm not going to pay that. 'Lucy,' I said, 'have you taken account of all the things I do for you?' In fact I couldn't think of a single one. In the end she kept me on 30 bob so long as I kept quiet and didn't tell the other lodgers – I was the only Bevin Boy. We were good pals, Lucy and I. I used to take her to the flicks.

Finding good alternative accommodation could take time. In Durham it took Brian Folkes a year and his move to the neighbouring village of Haswell was at the cost of another 5s a week – 'but at least I was away from South Hetton, which was blasted by a freezing wind off the North Sea in which vegetation cowered and died, with the notable exception of leeks that were cultivated with much tender loving care. And I again had a proper bed to sleep in.' Second lodgings were usually an improvement; Bevin Boys were cannier, or luckier. John Wiffen came down from the same hillside village as Denys Owen to live in Gresford itself, close to the pit, not because of the carnal desires of other Rhosllanerchrugog womenfolk but because he couldn't stand the bus journey and his breakfast of a boiled egg (a real egg – he was in farming country) wasn't enough to keep him going until snapping. He fetched up living with a farm worker and his wife, 'a truly loving, childless couple of an age that I might have been their son'. Denys Owen, Michael Edmonds and David Reekie developed the same close relationship with the couples who took them in the second time round: Owen in the home of an elderly ex-miner in the village of Pandy; Edmonds in that of the superintendent of his pithead baths; Reekie in that of a stoker at another colliery. Reekie was so happy in the tiny stone cottage high above Horse Riggs colliery that the lack of bathroom and hot water, and the communal toilets at the end of the yard which served three other families, seemed insignificant.

There were moments of embarrassment when one of the other women used an adjoining cubicle at the same time and wanted to talk – there was a single course of bricks between but it didn't meet the roof. I have never been a great conversationalist at such moments. I would mutter answers and read the newspaper that was to hand for an entirely different purpose.

For some men, seeking out new lodgings after a hard shift or a week of shifts was too much; resolve didn't translate into action. Alan Munford kept intending to look but didn't, partly, in truth, because he felt sorry for his landlady.

She always kept the front room locked, but one evening she felt ill and went to bed early, leaving the key in the door. We had a peek inside. There was nothing but a side table with an aspidistra, and a memorial clock on the mantelpiece with a brass plaque on it. The clock had been presented to her by Bentley colliery – the plaque told us her husband had been killed in the pit. The clock wasn't going. I wondered if she kept it at the time he'd been killed. No wonder she didn't have a clock that worked: time had stopped for her.

Munford did eventually go – after two years – but only because his fellow lodgers transferred nearer home and left him on his own. Things were much better in his new digs: he received the cheese ration to which he was entitled and got his laundry done, something that his first landlady never did, though it was supposed to be included in the rent. There was, however, a downside: 'When relatives came to stay they put me in the box room where I slept on planks supported on orange boxes.'

Men who alighted on good digs stuck with them, only moving if they had to. John Potts, who was at Shipley Woodside colliery in Ilkiston, Derbyshire, moved twice, unwillingly. 'All three lots of digs the three of us had were good but we got turfed out of the first two because the woman of the house got pregnant, nothing to do with us Bevin Boys let me hasten to add, and there wasn't room for us,' he says. 'The third was good too, except the daughter of the house suffered badly from constipation and would go for a fortnight and then occupy the outside privy for what seemed forever. We had to steer clear until the deed was done.' There was a variety of reasons for men to move on, whether it necessitated new lodgings or a move into a hostel, or out; sometimes, for instance, one or other could be within reasonable walking distance of the colliery, which made it an attractive proposition. When Ron Bown went ripping down the Florence pit in Stoke he couldn't get from his hostel at night because there wasn't any transport and he took lodgings that were closer. When Alan Gregory went packing at Cossall in Nottinghamshire, another night-shift job, he had to move from one lodgings to another – which, in fact, was right next door. The reason was that, in the first household, his landlord was a colliery stoker working days and his wife, who had two children, couldn't cope with men on different shifts; next door, the husband was a miner on nights.

Gregory was the most reluctant of movers. He'd come to the Uptons by chance. On arriving in Ilkeston, the nearest sizeable town to Cossall but on the other side of the Nottinghamshire border, he and another Bevin Boy had been directed to lodgings so run down and with such primitive facilities that they'd returned to the billeting office in the town hall to ask for something better. A council employee who had nothing to do with billeting, overhearing their conversation, suggested that a neighbour of his on the council estate near the cricket ground would be glad to help. 'And, by golly, did we fall on our feet,' says Gregory.

Charlie and Ada Upton were both kindly sensible people and their home was bright, clean and cheerful, if a little crowded. Ada was the most marvellous cook. George's and my ration books made some allowance for a manual occupation, but I was still amazed at the quality and quantity of food, in particular meat dishes, that we enjoyed. Derbyshire hotpot cooked slowly overnight in the oven of a coal-fired range comes fondly to mind. From my late teen to my early thirties my weight was steady at ten stone seven with one exception – the year or more of Ada's cooking when I was a miner. That year I went up to eleven stone six.

Harry Fowler, employed in Thoresby colliery in the north Nottinghamshire village of Edwinstowe, moved from lodgings to a hostel because his landlady gave him the spooks.

The room she gave me was right at the top of the house. She told me: 'Once a month you'll hear tapping on the roof in the middle of the night.' 'What's that?' I asked – I thought she meant it was a knocker-upper or something like that. But no. 'It's my husband,' she said. 'But I thought you told me he was killed down the pit.' 'Yes,' she said, 'but he comes once a month to tell me he's still thinking about me and he taps on the roof.'

*   *   *

For the Austrian Jewish refugee Meir Weiss, the Industrial North was as much a place of curiosity as it was to Southerners. 'The contrast with well-manicured Surrey was, shall we say, startling,' he observes. The town and cities were gritty with the tang of sulphur belching from the mills and the factories and the forests of domestic chimneys. (In Stoke, Geoff Baker says cheerfully, 'the early morning walk from my lodgings filled my lungs with the filth from the pot banks, steel works and slag heaps'.) In the grimy streets men and women went to work, many of them awakened by the window-tappers with their long poles: men in flat caps, women in curlers and shawls, men and women wearing clogs that click-clacked on the cobbles. If it wasn't for the bomb damage in evidence here and there (and until the end of the war brought the end of the blackout[4] and the return of the lamp lighters, who in the mornings put out the gas lamps), Northern England and tracts of the Midlands mightn't have moved on from their Victorian roots. 'It was,' says Doug Ayres, another who was at a Six Towns colliery, 'like living in a time warp'. That was true, too, of the small mining towns and villages built on coal and dominating the lives of those who lived above it.

In contrast to the bleakness of the many places they found themselves in, Bevin Boys on the whole received as warm a welcome in the wider communities as they did down the pits. Indeed, many received sympathy for being there at all. Men and women stopped them in the street to tell them or to say how much they appreciated what they were doing for the war effort. Some shops knocked a few pence off purchases. John Marshall, who before his nervous breakdown developed an obsession with never having enough soap, once went into a shop in Newcastle and produced his ration book but on hearing he was a Bevin Boy the woman serving him told him to put it away: "'My brother's a Bevin Boy," she said. "You can have as much as you want any time."' In the Derbyshire town of Alfreton, Reg Fisher and his pals at the nearby Ramcroft colliery in Nottinghamshire 'did very well from the girls in Woolworth's – free cups of tea and razorblades under the counter'.

The reception Bevin Boys received down the pits of south Wales was generally no different to anywhere else in the country; in the villages of the valleys around them, however, it was often considerably less welcoming. To some extent this was because, isolated as most of them were, they had little to do with each other, never mind outsiders, and they were suspicious and ill at ease with strangers in their midst. In Abergwynfi where he lodged, not far from Glyncorrwg pit where he served, Ernie Jefferies was aware that

> there was some resentment floating about over a bunch of London evacuee kids there whose parents didn't bother to keep contact – people thought they'd been lumbered. But none of the resentment was directed at us. I can't say the locals took us to their bosoms. There were only four of us and we stuck out like sore thumbs. Everyone knew everyone else. We were just strangers.

But in some other villages, Bevin Boys were treated with hostility. In Crumlin where he went to the Ocean colliery (renamed Deep Navigation after nationalisation in 1947), Dan White 'got stick from the women in the street. "My husband's off fighting for his country. My son's off fighting for his country. Why are you here?"' In Abertridwr where he served down the Lady Windsor, Albert Mitchell was frequently 'verbally assaulted. Their husbands were in the army so they reckoned we were taking their jobs. "I don't want your husband's job, get him back and he can have it," I used to say. They didn't get it. We were spat at by women many a time.' To Warwick Taylor, who was in Oakdale, it was understandable:

> This was the part of the country that had suffered the harshest deprivations of the twenties and thirties. The bitterness was deep. There was a fear that the bad old days could return after the war and perhaps we were the thin edge of the wedge sent to replace their kith and kin.

Whether in lodgings or hostels, initially Bevin Boys were too exhausted to think about anything other than work. Some fell asleep on the bus after a shift and had to ask to be given a shout when they came to their stop. Michael Edmonds repeatedly nodded off over his evening meal. Many in the hostels got back, took off their boots, lay on their bed until the evening meal was served, and awoke to find it was over. Almost everyone had aches and pains. Reg Taylor slept on the floor. Some slept from the end of the Saturday morning shift until they woke up on Monday and started all over again.

In time, of course, Bevin Boys became accustomed to the life: they were, after all, young men. And, like all young men, they wanted to get out and enjoy some sort of social life. For a good proportion that meant the miners' institute: subsidised beer and cigarettes a major pull for Bevin Boys nursing their pennies, though there was also a library. Most of the week the institute, like all workingmen's clubs, was an all-male environment and 'so smoky you couldn't see – nearly everyone smoked then,' says Warwick Taylor. 'I was a pipe man – a pipe was a bit of a status symbol.' Women were allowed in for the Saturday night dance (and in Oakdale,[5] where Taylor was, to the cinema that was part of the premises, the only one in the village). In an age with a less demanding idea of what constituted entertainment, many Bevin Boys joined the local youth club, 'a wind-up gramophone, darts and what we knew as ping-pong was the high life' – Stan Payne's recall. He enjoyed the Saturday night hop in Welbeck's village hall,

> where the dancing carried on for half an hour or so after the institute closed and some of the boozers came in for the final whirl – and even Fred the Cap had the last waltz with his misses. But he still never took his cap off. In three years there was never any bad behaviour or upset. It was clean, honest fun throughout.

In Stoke, Doug Ayres enjoyed a good, cheap evening out in the War Workers' Club,

> where Bevin Boys and other young miners, railway, pottery and shop workers came to drink tea, listen infrequently to lectures and dance. Members brought the latest records. It was the Swing era with Harry James, Charlie Barnet, Artie Shaw, Woody Herman and the inevitable Glen Miller. There was also a new young singer called Frank Sinatra.

Many Bevin Boys became mates with miners and went to the pubs with them to play dominoes or cribbage and have a drink, though one or two was all it usually was. 'We didn't have the money or the inclination to keep up with the miners,' says Alan Gregory. 'A Saturday night was enough.' It was in miners' company that more than a few Bevin Boys got their introduction to pubs and beer, in Norman Brickell's case Barnsley Brewery's Old Beer. 'I was never drunk,' he says. 'But I regularly walked up the steep Lund hill in Royston through the ever present chemical fumes from the pit coke ovens, pleasantly happy, ready for the early morning shift and the winding man's elastic revenge.' Many Bevin Boys threw themselves whole-heartedly into as many aspects of what life had to offer as they could. They played cricket, rugby and football on colliery teams or came to games as supporters. They went to sporting events with miners – Bill Hitch in Durham 'got taken to all-in wrestling in Newcastle and to watch Newcastle United and Sunderland', while across the Pennines Donald Whittle went with the men from his colliery to watch Bolton Wanderers. 'We shared a passion for football. On home match days there was always a rush to get out of the pit in time to get to the game where Nat Lofthouse, a fellow Bevin Boy, would be playing.'[6] Just as some Bevin Boys visited injured miners in hospital, so miners repaid the kindness. When Dave Moody was in the Newcastle infirmary with a fractured leg, some of the miners from Craghead colliery – on the way to watch Newcastle at home – came to bring him cigarettes.

Bevin Boys who made it clear that they wanted to belong were appreciated and accepted and were never short of invitations to weddings or christenings or a meal at weekends. Stan Payne felt free to visit other Bevin Boys who lodged with miners 'to spend the evening playing cards or just chin-wagging with the family'. To many Southerners the hospitality that the majority of miners extended once they'd got to know them took getting used to. 'Yorkshire folk were so very different to the average Londoner,' says David Reekie.

Back home I would have been hard pressed to tell you the names of neighbours other than those immediately over the fence. I was never invited beyond the front door. Insularity was inbred in me. At first I didn't understand all this 'Come in, lad, and mek thee-sen at home.'

It surprised many Bevin Boys to see the men they worked with in a different light, to appreciate how so many loved their gardens or allotments, their whippets and their pigeons – many a miner, as Reg Taylor discovered, 'had a coop in the yard, furnished with an old chair and some means of brewing up'. In the culture of the time, miners' wives didn't expect their husbands to help with household chores or with the children and those who visited their homes thought their lives were hard. Most miners' homes were low rows of old rented pit property (or pit property taken over by the local authority), many substandard, suffering from subsidence and difficult to keep clean. The fire in the kitchen range on which the women cooked and boiled all their hot water never went out, even in summer, when those in the house sweltered. In many Bevin Boys' memories, miners' wives seemed almost always to be washing clothes or bed linen, pummelling a scrubbing board, their sleeves rolled up. Many didn't even have a mangle. If the weather didn't allow them to 'peg out', damp washing festooned the kitchen around them; if it was, even on the finest days, as Reg Taylor remembers, 'the eternal dust found itself out of the local collieries and settled on window ledges and fences – and the washing hung on the line'.

At least miners' wives had a plentiful supply of concessionary coal to keep the kitchen range going, a precious benefit during the war when all other households experienced severe shortage. When Alan Gregory was living with the Uptons in Ilkiston, Derbyshire, there was more coal than enough and he and his fellow lodger George 'were happy from time to time to wheel a covered barrow to the homes of Ada's two brothers, who kept butcher' shops'.

The right to concessionary coal stemmed from the early thirties when collieries, unable to compete with continental prices, struggled to stay in existence and squeezed miners' wages to pitifully low levels. Rather than meet demands for increased pay, owners reached agreements to provide mining households with coal at a reduced rate. The basis on which this was allocated gave the miners less of a bargain than the prices they paid indicated. In the Midlands, for instance, weighbridges were adjusted so that they took 20½ cwt to register a ton – the half going into the concessionary coal pool. In County Durham, where coal was calculated by the score of tubs filled, a score became 21.

The agreement on how much coal men were entitled to varied across the coalfields, but was about a ton a month. In Nottinghamshire, miners paid 6s 9d for 23 cwt (which had a commercial value of about 45s) the year round; in Kent they paid 6s for 16 cwt every four weeks in winter and six in summer. Some pits charged cartage, others didn't. Either way the coal was delivered unbagged and shot into the street, from where it had to be 'got in' – which in the terraced rows meant through the hall to the yard. If they had no one else to help them, wives of miners away in the forces, and miners' widows, had to do the shovelling: no easy task as the coal was unscreened, just as it came up from the pit, often in huge lumps or mixed with stone. Lads could earn 6d for carrying out the task. As Brian Folkes remembers, the discarded stone was left in little piles in the street – hazards, until the end of the blackout, into which more than one Bevin Boy cycled on his way to work.

Who was entitled to concessionary coal was as variable as the entitlement. In some collieries everyone qualified if their home was within horse or lorry distance for delivery; the majority, however, restricted the right to married employees, while others restricted it to underground workers or even colliers. Some collieries were generous enough to consider that married Bevin Boys living in their catchment area qualified. Les Thomas at Yorkshire Main once helped an ex-service Bevin Boy who lived in Stockport get in his entitlement. '"Come up and help me shovel the stuff into the bunker and I'll make sure you have a good weekend," he said, so a couple of us went home with him. As it happens, my mother always had coal. The coalman told her: "I know your son is digging it out. I'll see you never go short."'

A few collieries were even more generous, extending the concession to the widowed mothers of Bevin Boys; Ron Bown's mum in Wolverhampton was a recipient. A very few collieries, notably in County Durham, extended the concession to any Bevin Boy in lodgings. Such Bevin Boys had the option of selling their allowance back to the colliery for around 10s – a slight profit; those who perhaps more astutely opted for the coal had their landlady's gratitude and, presumably, her best attentions.

The cockwood that some Bevin Boys brought back from the pit to their landladies also made them popular.

One of the few perks in mining was being allowed to take the off-cuts from pit props for firewood – cockwood as the miners called these. Managements forbade the removal of pieces over a prescribed length: 6in in most pits, 9in in others. Some pits had an abundance of cockwood: as Ken Sadler witnessed down Blagdon in County Durham, miners weren't above sawing up pit props. Down the Tower in south Wales Desmond Edwards remembers, some men were even more blatant:

They carried chain saws down the pit, all safely coiled up when not in use – they resembled motorcycle chain with a sharp tooth at each link and a ring at each end for the insertion of a wooden baton. Two men would sit on their haunches facing each other, boots against the tree trunk between them alternately pulling and pushing. Then the chains would be coiled up again and out of sight.

When peace returned the first 'Norways' arrived – lovely straight uniform conifers. A surge in chain saw activity. But even this was not enough for some. A large proportion of the workforce were Merthyr men. One day one of their buses was stopped on the way home by the police on the top of the mountain just as it was about to pull in at a quiet lay-by. Apparently not a single passenger could offer any explanation as to why the entire gangway was obstructed by a magnificent example of Scandinavian forest growth.

In the innocence of youth, Bevin Boys initially had no idea of the derivation of 'cockwood', and some were covered in confusion when they found out, as David Day did when he used the expression in mixed company, 'causing the men to look sheepish and the women to chuckle and call me a naughty boy. I had not realised what might happen on the mat in front of the fire when a blaze of this purloined wood was emitting its seductive glow.'[7] Perhaps Bevin Boys in Scotland, where cockwood was known as 'pussywood', were quicker on the uptake; or perhaps not.

Not all Bevin Boys mixed with miners outside work. As a generalisation, Bevin Boys who had most to do with them were those in lodgings, those in the hostels the least.

According to a leaflet handed to new arrivals by the National Service Hostels Corporation, 'the Hostel should not be a self-centred independent unit but an addition to an old established community. Your work will bring you into contact with fresh people in new surroundings. These new associates will be glad to welcome you into the social life of their community'. The problem was that for many Bevin Boys their hostel was an independent unit isolated from their surroundings. The adequate or better material comforts that it provided, and the companionship within it, were enough to prevent many trying to form other links. As Reg Fisher, who was in the hostel in Alfreton, says, though humorously: 'The inmates bonded in a mutual hatred of the whole thing.' Less humorously David Day remembers how separated Bevin Boys in the Cannock Chase hostel felt from the world around them – and that the barrier was preserved on both sides. 'The parties involved just did not want to mix,' he observes. 'I never went into a miner's home or made friends with a miner outside working hours, nor did any of the other Bevin Boys of my acquaintance.'[8] Meir Weiss and his fellow Bevin Boys in the Stalybridge mini kibbutz didn't mix with miners either, but in later years he came to think that in human terms they missed out. 'As a group we were too self-contained, too serious,' he says. 'All our time was spent preparing for life to Palestine.'

> In my second year I became a kind of teacher in the Hashomer movement. Manchester had a large Jewish community and the immediate post-war years were the height of the Zionist youth movements. It meant moving to a bayit [house] in Prestwich, the only miner there, teaching 14-year-olds – a bit like scouting but with a little Jewish education thrown in. There was no time for other things.
>
> I regret not having more contact with other people, not getting to know miners' families – we Jews didn't even mix with other Bevin Boys. But we were totally absorbed in a vision of the future. And we were not in a real mining community, just part of a big city, really. It might have been different in a real mining community.

If they largely turned inward, Bevin Boys in the hostels heeded another paragraph in the hostels corporation's leaflet: 'In the Royal Navy they talk about a ship being a happy ship. You may remember it in the film "In Which We Serve". We want this Hostel to be a happy Hostel. We know that much depends on the staff and how they run the place, but a lot depends also on you.' Taking that to heart, Bevin Boys formed residents' committees concerned with welfare, sport and entertainment. Those with talent got together and put on shows – Reg Fisher, who from sixteen belonged to an amateur concert party that toured ack-ack batteries and barrage balloon and searchlight bases, formed a group, a pianist, a drummer and himself on double bass, with the vocals done 'by two lads we heard singing in harmony in the showers'.

The Alfreton hostel had other talents: Max Semel, a professional singer called up from ENSA (who as Paul Vaughan later sang with Ronnie Scott's band), and Reg Powell, the Welsh amateur lightweight boxing champion – whom Fisher got 'punching a ball to music with his silhouette thrown up on a sheet'. All the hostels ran dances. In Alfreton, Fisher 'scrounged two turntables and a PA set from somewhere and the lads brought their record collections. I was a DJ before they had DJs. The girls came to our Sunday hops from miles around. They had to tell their parents they were visiting friends – Sunday entertainment was frowned on.'

On the whole, hostels were happy, though those who preferred to keep themselves to themselves found the noise and the constant comings and goings unsettling (and drove some out to find lodgings). With men in the same huts on different shifts, sleep was a problem, especially in the day; another reason why only the morning shift was popular. But the majority of men adjusted to communal life. Many changed hut to be with like-minded others, Geoffrey Mockford in the Abbotts Road hostel in Mansfield moving in with 'three public schoolboys, an ardent communist, three keen motorcyclists and the rest people like me with ordinary jobs, but stimulating'.

In training, Ernest Bevin sent all recruits at the training centres a letter urging those already embarked on a course of study for their profession or occupation to continue with it, and promising those who wanted to undertake some other study that they'd get every assistance from the education authorities. Bevin Boys, his letter pointed out, were going to the mines to help produce more coal and

> it would be unreasonable to expect that you would be allowed time off during working hours to attend classes, unless, perhaps, you were studying mining with a view to making it your career, but evening classes in the towns cater for most of the popular needs and some of the more specialised subjects as well.

Hostel conditions weren't conducive to study but Mockford, convinced 'that education was the way to make progress in life after mining', began a correspondence course for a School Certificate and stuck at it for the necessary two years. Few Bevin Boys around him bothered to take advantage of the Minister's urging: out of 350 Abbotts Road Bevin Boys, 'no more than the same five or six of us ever used the small reading room. One of the things that struck me was that there were always bodies all over the place, snoring.' It's unlikely the percentage that studied was greater in other hostels; it took a lot for a young man to bother with books after the physical labour of a shift in the pit. It was as much as most could do to write a letter home; as Eric Gulliver found, writing his Mass Observation reports, a day's shovelling made his hand shaky.

It was easier for men who wanted qualifications to knuckle down where they had some peace and quiet: in their homes or lodgings. Peter Archer did so at home in Staffordshire (carrying on with the BA correspondence he'd begun while awaiting call-up) as Ian McInnes did at home in Kent and Victor Simons in lodgings in Derbyshire.

Simons, who'd volunteered to become a Bevin Boy with the intention of making mining a career, got the day release the Minister promised to attend the mining department at Chesterfield technical college and then, in the middle of 1946, began a Sheffield University BA Engineering degree, permitted to do the first year externally at Chesterfield. As the war in Europe ended, McInnes, deciding on a career in mining engineering, began an ordinary National Certificate at Ashford tech, realising how lucky he was:

> A day off a week to go to college, with full pay for the day provided I attended one evening class a week. I was just turned 20 years of age, working a four-day week with three days up in the fresh air and sunshine, getting paid £3 a week, living at home and enjoying my mother's cooking.

Other men already on a career path added to their qualifications. In Stoke, Doug Ayres took an Intermediate BSc in inorganic chemistry. In Durham, Brian Folkes swotted for the Royal Institute of Chartered Surveyors' first exam and was granted leave to take it in the spring of 1945. 'It was entirely at the colliery manager's discretion and I appreciated his decency,' he says. 'That wasn't something I associated with the Ministry of Fuel.' Ian McInnes found the Ministry sympathetic in the bad winter of 1947 when he applied for his release four months early so that he could begin a mining degree at Nottingham University – this at a time when coal stocks were critical and the recently formed National Coal Board was pleading with Bevin Boys coming up for release to stay on temporarily. 'I swung it by pointing out I wasn't leaving the industry,' McInnes says. 'In fact, when I retired, my demob was 41 years late.'

Bevin Boys living at home had the advantage of being able to pick up the threads of their previous existence. Ian McInnes's sports were rugby and rowing and 'as most of the experienced players and competitors in the district were in the forces I was able to play three-quarter wing for Dover for several seasons and in the summer row for Dover in the coxless fours at a number of south-coast regattas'. Transferring to a Midland collieries, Ron Bown got back to the Boys' Brigade in Wolverhampton and Ronald Griffin to acting with the YMCA Players in Smethwick, appearing in plays and revues for various war charities in and around Birmingham. Down the pit he rehearsed his parts, his lines bouncing off the walls. Once, when he was to appear as Macbeth, he was declaiming 'Is this a dagger I see before me, when I became aware of a pair of eyes staring at me, this little miner said I wasn't right in the head and scurried off. That got me the reputation as "the chap as spakes to hiself."'

Although he had good lodgings, as the only Bevin Boy in Horse Riggs pit David Reekie initially 'felt out on a limb'. He 'whiled away Sundays in solitary confinement, and evenings walking and envying the sounds of family life drifting from the houses I passed'. Finally he gravitated to the chapel, 'for the music, not

out of any religious fervour – and singing changed my whole life as a Bevin Boy.'
Possessor of a fine bass-baritone, Reekie had performed in amateur musicals at
home; here he soon had a leading role in the chapel's annual presentation of
The Messiah (fee three guineas) and was singing at fund-raising events at other
chapels in the district – supported by miners from the pit. Forming a double act
with a local girl, he toured army camps, airfield and prison, finishing up 'with the
dubious reputation of having been in every prison in Yorkshire'.

Most Bevin Boys with a taste for culture managed to satisfy it even on their pay.
In Sheffield, Tony Brown had a good rep, which put on Priestley, Shaw and O'Casey,
occasional ballet at the opera house and symphony concerts with Barbirolli and
the Hallé Orchestra, also seen by Geoffrey Mockford in Mansfield and Doug Ayres
in Stoke.[9] In the Pottery towns and adjacent Newcastle-under-Lyme, Ayres 'had
some 50 films a week within a 2d bus ride. Some were pretty ancient, old films
from the 1920s onwards, plus obscure B movies. I remember being impressed by
Erich Von Stroheim.' Tony Brown sometimes booked a seat at a cinema in Sheffield,
but preferred 'the bug-hutch' near the hostel in Woodhouse, 'which put on films
only after general release but had the advantage of double seats upstairs, ideal for
necking with the current local girl'.

Girls, naturally enough, occupied young men's minds and the majority of Bevin
Boys went to the dancehalls in pursuit of them, the shy and those unable to dance
usually doing nothing more than watch, and yearn: for all Bevin Boys the seams
that women ingeniously pencilled on the back of their legs in some semblance of
the stockings that weren't available to buy in the war held greater appeal than
those that dominated their working lives. At the start of the war, dancehalls, like
cinemas, had been closed but soon reopened and were packed everywhere. The
public's determination to be cheerful in adversity made dancing livelier: quick
steps became more popular than waltzes and 'excuse mes' and novelty dances
enlivened the floor. The greatest innovation was jitterbugging, which arrived with
the Americans in 1942.

Alan Lane spent more time than most in the two dancehalls that Pontypridd
had to offer, 'got the dancing bug' (in his eighties, he still has it) and, proving nifty
on his feet, was less than popular with rival young locals. John Cook, another
dancehall regular at the hops held at the miners' institute in New Kyo, at the
Castles ballroom in Annfield Plain or the Palais, the Hibernian and the Co-op Hall
in Stanley, was so successful with the local girls that he had 'some serious feuds
as did other of the lads – the hostel itself was attacked a number of times before
things settled down'. 'There were often scuffles with other young blokes, those not
yet called up and some younger miners,' recalls Phil Robinson, another in County
Durham, at a different hostel.

But we had quite a few Glaswegians among us and, by gum, they would
sort anyone out. Actually we had more run-ins with the army – squaddies

up from Catterick. We got into a terrible fight in Durham at one of the pubs. There were about 20 of us Bevin Boys and we took them on. They got a right hiding.

The problem for most Bevin Boys was that they were impecunious, sometimes without enough to pay for admission to a dancehall, never mind buy a girl a drink. A group from the Alfreton hostel, Reg Fisher recalls, once went to Chesterfield when they were broke,

but we had Harry Fowler with us that night – he used to come over regularly from his hostel to see some of the lads he'd been in training with in Creswell. Anyway, this dance was in a place over Burton's. Harry kept the bloke on the door chatting while the rest of us slipped in behind him. Harry could talk a bird out of a tree.

Even when, in their terms, they were flush, Bevin Boys were among the also-rans in amorous matters when the Americans, who seemed to be stationed everywhere, were on the razzle. They had money to burn, as well as an aura of glamour – and enviable uniforms. But they were generous to many Bevin Boys, standing them a pint while commiserating with their lot. When he was in training in Nuneaton, the Americans at a nearby camp invited Roland Garratt's entire hostel to a dance, even laying on a coach. 'It was a smashing evening with the drink free flowing. But then one of the Yanks devoted too much attention to the girlfriend of a local Bevin Boy and things soon developed into a brawl. When the American military police arrived the Yanks were escorted back to camp.'

How men brought variety into their lives outside the pit was a question of interests, temperament and opportunity. Victor Simons joined a local amateur dramatic society; Alan Gregory resumed piano lessons, 'though I had absolutely no natural talent'. Geoffrey Mockford largely lived in the public library; so did Norman Brickell, though the reading matter was less important than 'the attractive young librarian, the object of silent adoration from this callow young man who lacked the courage to ask her name'. Brian Folkes spent most weekends at Durham ice rink, 'which had a canvas roof until, one typical winter's night, it took off and was last seen heading out into the North Sea'; John Wiffen worked on the farm where his landlord was employed, 'not for any reward, that would have spoiled the pleasure – though the farmer's wife sometimes had a basket for me to take home at weekends, containing eggs and a chicken'.

In the annual week's summer shutdown, and after nationalisation brought the five-day week giving them clear weekends, almost all Bevin Boys escaped from the mining areas into the countryside surrounding them: the Dales in Derbyshire and Yorkshire, the Peak District, the Lake District, other areas that were untouched by the war; Len Ungate cycled hundreds of miles through the

Welsh valleys. Many others brought their bikes from home; others used the buses, which were cheap and, even when the war was on, surprisingly regular. While he was in Durham, Brian Folkes explored from Whitby to Edinburgh, staying in youth hostels (as many other Bevin Boys did), at a cost of about a shilling a night, 'ideal for impoverished young men'; he continued hostelling when he transferred to Kent where, 'in the magnificent summer of 1947, I was able to take my fiancée with me'.

Some Bevin Boys were engaged before they were called up. One was Bill Gibbs, who almost every weekend biked from his colliery near Mansfield to his home in Grimsby (where his fiancée was a hairdresser) – a trip of 65 miles in each direction. 'Coming back I'd leave Grimsby after midnight on a Sunday, ride through the night, have my breakfast in the pit canteen and go down the pit,' he says. 'Now Joyce says I'd have trouble cycling to the next corner for her. Sad but true.'

Other Bevin Boys became engaged to girls they met while they were in the pits, including Ron Bown. 'Who doesn't believe in fate?' he says. David Reekie, having avoided his landlady's well-intentioned attempts to pair him off with the farmer's daughter who delivered the milk, became engaged to the first girl he danced with in Morley town hall, 'despite her being unimpressed with my prowess – I had a rooted inability to perform jive and Latin American numbers'. George Ralston in Newtongrange and Morry Pearce in Doncaster both met the girls they would marry on VJ Day, which truly marked the end of the war; it was also, propitiously, Ralston's nineteenth birthday. Pearce had frequently seen 'this lovely girl with the super long legs' as they cycled passed each other, her coming from work, him on his way for the afternoon shift at Bentley colliery, but they'd never exchanged a word until they both won prizes at the same street party. He was smitten. And when 'someone asked for someone to carry a small bath of apples to the local hall for bobbing for apples and Sylvia took hold of one end of it, I was on the other end before anyone could move'.

At least a few Bevin Boys married while they still served in the pits, including Bill Hitch in Durham, who married a girl he met at the cinema in Sacriston, and John Cook, who fell for a girl on the staff at his Durham hostel in Annfield Plain – and another Bevin Boy married her sister. Most, however, like Geoff Baker, waited until their time was done. Cricket, indirectly, brought him and his future wife together. 'It was after an evening match in Stoke and I got talking to one of the other players,' he says.

His name was Billy Briscoe, who played football for Port Vale for many years and cricket for Staffordshire. At this time he kept the Black Boy pub in Corbridge. Billy invited me home for supper, and that's when I met his daughter and that, really, was it. As it happened I was seeing another girl in the same street, so she had to be pensioned off.

Baker and Marjorie were married by his father, who'd moved from Dockland to St Peter's in Broadstairs, his home town. When it came to it, Baker didn't follow his father into the ministry as he'd told Big Joe and the other miners he'd worked with he expected to do. 'I was seen by the Advisory Council and asked about myself, what I'd done in my leisure time as a Bevin Boy,' he explains.

'Well,' I told them, 'I went to church on Sunday morning, but then I'd go to the pub at lunchtime for a drink with Big Joe and Elsie his wife or with other of my miner friends, and I'd play cards and cribbage with them; and during the week I'd go to dog racing with them. Undoubtedly they were the wrong answers and I was turned down as being too worldly. I'm sure the Council were right in their decision. Mining had changed me – for the better, I think. I had no regrets.

Eight

On the Black Side

Ernest Bevin sent in his Boys to fill the gap in the miners' ranks and to help provide the coal for the final big push of the war. In 1943 output had dropped to 194 million tons. There were hopes that in 1944 the Bevin Boys could make a difference of around 7 million tons (even 20 million tons in 1945). What happened was that output dropped another 10 million.

Drafting the Bevin Boys to bring the workforce up to something like the number deemed necessary did nothing to reduce the strikes and absenteeism that had dogged coalmining after the first eighteen months or so of the war. In fact, strikes in 1944 escalated alarmingly. And the worst of them – the biggest dispute in the pits since the 1926 General Strike – erupted in January even as the first intake of conscript miners was in training. The cause was the Porter pay award. And the government was stunned by the miners' reaction to it, which not only gave them their third increase of the war but the highest minimum wage in the country.

The miners' leaders had sought a rise to £6 underground (from £4 3s) and £5 10s on the surface (from £3 18s). Porter gave them £5 and £4 10s (and coincidentally raised the Bevin Boys' money after they'd protested they couldn't pay their way on what they were getting and walked out of the training centres). Within twenty-four hours almost every pit in south Wales and Lancashire was idle and soon joined by pits in Yorkshire, Kent, Scotland and Durham.

It was less the pay award itself that caused massive anger among the miners but that the Porter tribunal had turned down a claim for an increase in piece rates. The award therefore wiped out the cherished differentials of colliers and other higher grades. In Lancashire, the area with the top rates of pay, the award represented only a very small increase for the majority and nothing at all for the skilled.

Over the next two months the Minerworkers' Federation tried to get the district unions to overhaul their complicated wage systems and agree some uniformity, which proved a slow and frustrating business. Many pits did go back, grumbling. In County Durham (and to a lesser extent elsewhere) men began 'ca'canny' working – working to rule, which in Durham alone reduced output by 50,000 tons a week. In Yorkshire, however, 120,000 men stayed out. Almost in despair the government served 400 of the 2,500 men at Easington, which was at the centre of the strike, with fourteen days' notice that they'd be moved to other collieries, and threatened to close the pit. At the end of March Lloyd George, the Minister of Fuel (whose department was having to reroute as many Bevin Boys as they could away from the trouble spots) hammered out a new deal. Effectively, he restored the differentials, in return winning an agreement from the Federation that pay and piecework rates would stand for four years.

That should have been the end of it, but it wasn't. The colliery owners were less than happy. The Porter award was to cost them £5 million a year and they expected the government to meet the additional £7–9 million that the Lloyd George deal added to the bill. The government refused. Whether in retaliation or not, some owners in Yorkshire marginally increased the cost of home coal to some of their employees: and seventy-two pits that had briefly returned, downed tools.

At a lunch on 4 April, Bevin, outraged, described 'what has happened this week in Yorkshire is worse than if Hitler had bombed Sheffield and cut our communications' and went on to say: 'Whether I shall survive, and my policy [of relying on industrial negotiation and arbitration], or whether other steps will have to be taken, I cannot prognosticate.' For a change Bevin was being deliberately oblique rather than inarticulate, but the message was clear: the big push was looming (D-Day was two months away) and the government wouldn't hesitate to put mining under military control.

There was no support for the Yorkshire miners. The TUC accused them of having 'struck a blow in the back of their comrades fighting in the armed forces'. The Federation and even the *Daily Worker* urged them to go back, which they duly did, a week later. In the wake of the strike, Bevin introduced an amendment to the defence regulations, which made incitement to strike unlawful across all industry. The penalty was a maximum of five years' penal servitude or a fine of £500 or both.

Between January and the end of the Porter furore nearly 900,000 working days were lost in the pits.

Britain at war was united in a common resolve but, contrary to popular mythology, was frequently disunited. More than 1,000 disputes occurred almost every month. Most were brief, all were illegal under wartime legislation and in 1944 they accounted for a collective peak of 3¼ million lost days.[1] The strains of war and, ultimately, war weariness were underlying causes. But there was another factor: a new mood of resistance against employees, a consciousness that rank-and-file organisation backed by union power put workers in a strong bargaining position. And nowhere was the trend towards industrial action more evident than in mining. In 1943 it accounted for half the days lost in Britain; in 1944, when there were more strikes in the pits than in any year since the beginning of the century, it accounted for two-thirds.

Wilful absenteeism, too, had a devastating impact on coal production. Sickness and accidents took their toll but in 1943, when 12 per cent of shifts weren't worked, two in three men had no excuse for staying off work. The worst offenders were the colliers, whose absence in mechanised pits destabilised the production cycle – not only was their contribution to output lost but as their stints remained unworked at loose it, the backshift men were prevented from moving the face forward for the next day.

Miners' leaders strongly condemned absenteeism. At the time of the 1942 pay rise Will Lawther, the president of the Federation, told the annual conference that the miners now had 'new duties and responsibilities'; three years later the man who replaced him, Arthur Horner, would talk about the need for 'a new morality'. A large minority weren't listening.

Bevin Boys' experience of strike action varied. Those in the Nottingham, Derbyshire and Staffordshire coalfields, which generally hit their production targets, found few if any labour problems. 'I witnessed some fairly colourful threats to strike,' says Peter Archer, 'but they were always settled.' There was little disruption in north Wales, though south Wales was responsible for a quarter of the plunge in output across Britain. 'No strikes in Gresford, it was a happy pit,' remembers Denys Owen. 'What was interesting to an outsider was that the north Wales miners didn't like the miners from the south. I recall a strike committee coming from the south by bus to persuade us to strike. They were sent away with a flea in their ear.'

Everywhere, however, Bevin Boys quickly realised how rife absenteeism was. The Sunday night and Monday morning shifts were always the worst. 'Oh dear,' says Alan Brailsford. 'I don't think I ever worked with a full team after a weekend. I think they used to think they didn't want to be killed with any money left.' Observes Ian McInnes:

Many didn't turn in after a liquid weekend. And if it was a sunny summer Monday, then, wage packet in hand, they'd just say sod it. Bevin Boys sometimes benefited if there weren't enough men to work the faces. The underground manager would strike a deal. If you filled, say, 200 tubs he'd let you go up and pay for the full shift. On occasions like that we worked like bloody maniacs – sparks flew from the tub wheels.

It amused Meir Weiss that whenever there were mid-week football fixtures or horse or dog meetings, miners' absenteeism could reach Monday levels. 'We used to pull their leg,' he recalls. '"Been playing again?" we used to ask.' Desmond Edwards remembers three miners being so anxious to get out of the pit to go to the races that they resorted to elaborate subterfuge:

The overman was going his rounds and saw two men struggling with a third, who was foaming at the mouth and displaying the signs of epilepsy.

With great difficulty they got him to the pit bottom where he broke away and attempted to jump in the sump. He was dragged away and all three were wound to the surface. Here he broke free again. The two set off in hot pursuit, caught up – and all three jumped on a motorbike and sidecar and off they went to Chepstow.

Many Bevin Boys found it paradoxical that men who so casually took time off when they felt like it toiled so unflaggingly when they were at work – and how intolerant they were of any Bevin Boy who tried to take it easy. 'The minute that a Bevin Boy showed how well he could slack, he was exposed to ridicule,' wrote John Platts-Mills.[2]

It was at first careful and moderate, but soon developed into real hostility. If a Bevin Boy persisted in idleness for longer than a week, the miners would get him shifted and he would end up working by himself or working in one of the rottenest jobs in the pit.

'Miners were strange men,' says Denys Owen.

You couldn't really rationalise their attitude. They were patriotic but fiercely independent. The older guys were used to austerity and didn't seem to want anything better. None of them was interested in saving for the future. There was nothing to spend their money on, other than drink and betting and there was a limit to that. They had enough for their needs. So for a bit of leisure they were prepared to lose a shift here and there.

* * *

The miners were hardly role models, but absenteeism among Bevin Boys was at first rare, 'not just because we were all broke, but because we felt we were doing our bit,' as Denys Owen sums up.[3] After D-Day and the real certainty that the war would be won, at which time the first conscript intakes had been in the pits for five months, commitment began to weaken. Absenteeism spread. A government survey five months on in November showed that wilful absenteeism in the Bevin Boys' ranks was half as much again as the miners – three-quarters as much again in Scotland.

The government expressed disappointment and told the Bevin Boys to pull their socks up. Christmas came and absenteeism reached new heights – and the government had no one to blame but itself. Once again it hadn't thought ahead.

Traditionally the collieries shut down only for Christmas Day. But that meant that Bevin Boys who travelled any distance home would have had to come back virtually as soon as they stepped inside their front door in order to be down the pit on Boxing Day. Most in that position chose to stay for the holiday and return to work three days late; many stayed away until after the New Year.

Just called up and on his way to Durham at the beginning of January 1945, Derrick Warren arrived at King's Cross to find

> not just other new Bevin Boys but God know how many others – together we filled up two coaches and some of us were sitting on our cases in the corridor. This load of chaps from the Durham pits had been home for Christmas and taken the New Year off as well, and they knew this was the train back.

They were a bolshie crew, Warren remembers, angry that they'd had to pay their own fare – Bevin Boys, he learnt, were entitled like all transferred industrial workers to two travel warrants a year at a cost of 7s 6d but could use them only between April and September – and rebellious that they now faced punishment for absenteeism under the Essential Work Order. Warren also remembers that

> the railway people locked us in. Maybe they thought some of them were so bolshie they'd do damage or maybe do a runner, though that was ridiculous, if they'd had that in mind they wouldn't have been on the train. There were no toilets in our two carriages. Hours later we stopped at Eaglescliffe, Stockton on Tees. There was a war factory of some kind by the station. One of the lads called out 'They'll have toilet and a canteen' so everybody climbed out the door windows and in, until we were chucked out by security. The train was stuck there until we piled back through the windows.

Their absenteeism cost hundreds of Bevin Boys across the coalfields a fine of £2 on each of three summonses. 'We went to the court, which was in Durham city jail, from pits all over County Durham – 118 of us the day I was there,' says Bill Hitch, who'd been home to Cambridgeshire.

> The manager said, 'I'll speak for you lads from my colliery', and he did. 'These lads are good workers,' he told the magistrates, but it didn't make a hap'orth of difference. I told them I couldn't pay, but my mother sent the money and she couldn't afford it either, but she didn't want me in the nick.

For the same reason most families found the money for the fine, including John Cook's and Dave Moody's; they'd both been home to Hampshire. 'I was prepared to go to prison because I hadn't the money,' says Moody. 'My dad wouldn't let me. "If you go down that line," he said, "it'll stick to you for the rest of your life."' A few Bevin Boys did serve twenty-eight days, returning to the pits with cropped heads. And when they did, some collieries imposed another fine – as much as

£5 – for absenteeism while inside. 'That,' retorts Bill Hitch, 'was diabolical. It didn't happen to me, but I know it happened, absolutely, one of them was a good friend of mine.' Many more who wouldn't or couldn't pay got away with it – there simply weren't enough prison cells to hold them, so the authorities expediently forgot about the matter.

Considerably more absenteeism occurred among Bevin Boys in the hostels than those in lodgings or living at home. Being among others who weren't able to come to terms with what had happened to them had a tendency to engender a collective lack of enthusiasm for the pits, 'especially among disgruntled southerners,' according to Bill Gibbs, who lived in the hostel at Forest Town close to Clipstone colliery near Mansfield. The all-lads-together atmosphere in the livelier hostels also had a deleterious effect on some. For most of his service in Yorkshire, Norman Brickell lived with an aunt and uncle. When, however, the war ended and they treated themselves to a holiday, he moved into the hostel at Dodworth near Barnsley. It was meant to be a temporary measure but Brickell liked it so much he didn't move back. 'The trouble was I started to enjoy myself. I was happy – too happy. From being a reliable worker I became anything but. Hostels weren't conducive to work. They were too comfortable; they made absenteeism an easy option.' It was all too easy on a cold dark morning or when the rain was rattling on the tin Nissen roof to turn over and not bother going on shift. 'It became a common event at early morning call for a voice to say "I don't think I'll go in today",' says Geoffrey Mockford. 'This was fatal. He was soon followed by others. The phrase became a joke. If a Bevin Boy version of *Dad's Army* was ever made it would be one of its catchphrases.'

When Reg Fisher and the other inmates of the Alfreton hostel fancied a day off – without repercussions – they showed ingenuity.

One of the lads found out that if we didn't have a breakfast we were entitled to be excused work. We were up long before the manager of the hostel arose, so we got the girls on the cooking staff to report a power cut – a regular feature in those days. No breakfast, no shift. The manager took what he was told in good faith and reported it to the Ministry of Labour. We pulled this stunt every nine or ten days and they never twigged.

At times Bevin Boys were legitimately absent from the pits, signed off by a GP because of injury or illness: eye infections (dust), gastroenteritis (eating with dirty hands), chills or flu (working in constant draft at pit bottom or coming from the heat near the coalface and up into the cold) were common. But Bevin Boys discovered that some doctors were unexpectedly ready to issue a medical certificate; one of the first bits of information that new arrivals at a colliery got was which local doctor was the soft touch. In Derbyshire Geoffrey Mockford limped to the surgery of the one he'd been advised to register with after a lump of coal crashed on to his foot with enough force to distort the steel toecap, 'expecting to get a note for a few days off. Instead he told me I needed two weeks and to go home to recover. I didn't need two weeks but you don't argue with medical opinion.' Complaints didn't even need to be genuine, as Alan Lane became aware in Wales. 'A very obliging lady doctor in Pontypridd was sympathetic to the Bevin Boys and could always be relied on if you wanted a week off.' When Des Knipe in County Durham wanted a week off, he went 'to the doctor who was easy, not the other one who was a martinet and played it by the book.'

> I said I had lumbago – everything to do with backs was lumbago in those days, they didn't know about slipped discs. He said take a couple of weeks off. Being cheeky I asked 'Could you make that three?' The fare home to Kent was three quid. If you were away for three weeks you got a free ticket. He was a good bloke, did that for me a few times.

Bevin Boys' own GPs at home could be at least as accommodating. The colliery doctor in Accrington signed off Harold Gibson for a week with a hand injury caused by a carelessly thrown pit prop; his own doctor in Blackburn gave him a certificate for three, as did Geoff Rosling's in Bristol for a similar injury. 'The pit manager at Ynysddu wrote to Dr Lewis to ask why the treatment was taking so much longer than they thought it should. Dr Lewis replied in his inimitable Irish way. I'd have loved to see the letter – he was very anti the Bevin Boy scheme.' When Ronald Griffin was stabbed in the knee by a piece of jagged metal protruding from a fast-moving tub, his boot filled with blood, but for two weeks he carried on working; then the wound festered and his GP in Smethwick ordered him out until it healed. At the time Birmingham Rep were staging a production of *Timon of Athens* (as a parody of wartime London) and looking for 'walk-ons'. Griffin auditioned.

> By the third week of the play my knee had completely healed but the kind doctor kept issuing me with notes – at a shilling a time. Finally I was ordered to be examined by the colliery doctor, who despatched me forthwith to the pit. Luckily the play had just finished its run.

In early 1944, the government removed numbers of Bevin Boys who completely refused to cooperate and posted them off to the army – but that encouraged others to copy their behaviour in the hope of getting away from the pits. Mostly, the collieries dealt with the Bevin Boys in their employ. Initially most tried to get those who made a habit of absenting themselves to mend their ways by putting them on unpopular jobs or shifts. Some moved men out to other collieries. And in conjunction with the government they fined offenders – as a rule of thumb, those who missed more than two shifts in a month. Pit committees, made up of an officer from the Ministry of Fuel, someone from management and a union representative from the workforce, were introduced in the middle of 1943 to deal with cases of miners' absence.[4] Now Bevin Boys stood in front of these committees and were fined £1, to be donated to a charity of the local union's choosing. If, however, they maintained a full worksheet for the six weeks after (if, as Les Thomas puts it, 'you were a good boy'), they got their money back. It was hardly a punitive measure and even less of one on the frequent occasions that one or other of the colliery members couldn't be spared and the official saw men on his own. As often as not, it seems, he was an avuncular fellow who popped into the colliery, borrowed the manager's office, sat in front of the fire smoking his pipe – and instead of a fine issued a warning, perhaps more than once.

*   *   *

Millions were directed into unfamiliar and unpleasant jobs during the war. Men who weren't conscripted, and by 1943 90 per cent of women under forty, were sent into munitions factories to fill shells, into tank and aircraft factories, steel works, chemical plants and elsewhere. Tens of thousands went to other parts of the country (dockers from Southampton to Clydeside, for instance), many putting up with poor working conditions and inadequate accommodation, none faring worse than the 90,000 volunteers of the Women's Land Army.[5] Yet to some people it seemed that the newspapers and the politicians were concerned only with the deprivations of the Bevin Boys. After the Bevin Boy strike at the beginning of 1944 had resulted in a pay increase, a John Purves of Exeter wrote to *The Times*, asking MPs

> to extend their solicitude to a harder case still – a case which is voiceless and still uncomplaining . . . I refer to those girls of 18 and 19, large numbers of whom have been sent away from home, often from very great distances . . . in order to make munitions. The pay of these young women is 42s 6d a week (or 45s at 19) and the cost of board and lodgings has in most of the areas gone up to 30s a week, thus leaving only 12s 6d a week to pay for bus and tram fares, midday meals, health and unemployment insurance, income-tax of 2s 5d a week, clothing, boots and shoes, and all incidentals.

It was never the harsh conditions they experienced that made the conscript Bevin Boys themselves consider that they stood apart from the rest of the civilian population: it was the feeling that they were soldiers or airmen or sailors denied the right to fight for their country and then denied the parity with the armed forces that they might have been granted. Servicemen based in Britain had three paid leaves a year and got additional forty-eight-hour passes; Bevin Boys' entitlement was one week, and Christmas Day. Servicemen who were ill continued to draw their money; Bevin Boys didn't, getting only 3s 4d sickness pay a day to a maximum of £1 a week, unless there was a top-up from the union – which didn't even cover bed and board, which to servicemen was free. There was no lifelong compensation for the dependants of Bevin Boys killed in the pits, as there were for those of servicemen who lost their lives. And no war pension for those like Ken Tyers, invalided out after sustaining a broken pelvis and ruptured urethra, which kept him in hospital for a year. Tyers did try to make a claim. The War Pensions Agency turned it down, explaining that he was eligible neither as a conscript nor as a civilian. As a Bevin Boy he had been 'called up and allotted a National Service Registration number [but] was not enlisted and therefore remained a civilian'. As a civilian his 'physical injury [was] not caused either by the enemy or in combating the enemy' and therefore didn't qualify either.

During 1944, many conscript Bevin Boys felt taunted by being expected, as civilians, to continue serving in the Home Guard, which happened in areas where local units weren't up to strength. Roy Eddolls, a garage mechanic from Blechingley, Surrey, who'd been in the 9th battalion West Surreys at home and transferred to the 21st King's Own Yorkshire Light Infantry, enjoyed the parades, which were held at Goldthorpe colliery; he remembers the last, when the Home Guard was disbanded in the autumn. 'We marched all around Goldthorpe village led by the Hickleton colliery band. Marching to a band was marvellous.' Others felt obliged to report for duty but, like John Potts in Ilkeston, Derbyshire, resented it.

> It seemed wrong, I was already doing my national service. It also seemed ludicrous at that period of the war. But I turned up a couple of times a week. I have memories of Sunday mornings crawling about on the grass in the park on manoeuvres, mainly trying to avoid the dog dirt.

Other Bevin Boys flatly refused, including Geoffrey Mockford who, just before the war in Europe ended, received a letter telling him to report to the Home Guard headquarters in Mansfield. He was the only inmate of the Abbott Road hostel to get such a communication. As instructed, he went on the Sunday – but only to tell the commanding officer ('a typical army officer of World War One vintage, with a waxed moustache which stood like thin black daggers either side of his nose') that he wasn't having any of it. 'I told him I'd been manipulated once already and I wasn't going to let it happen again,' he says.

Growing heated, the officer said Mockford would do as he was told and insisted on him being issued with full kit. Mockford took it, left, and heard no more until after the Home Guard was dispersed, when he received another letter, this one thanking him for his service and seeking the return of his equipment. He was allowed to keep the overcoat. 'And very grateful I was in the very cold winter of 1946–47.'

A constant painful reminder of Bevin Boys' lack of military status was that they were barred from British Legions and NAAFI clubs or canteens. When Warwick Taylor contracted pneumonia down Oakdale pit in south Wales and was sent to recover working in the open air on a gun site near London, he was able to use the NAAFI there, which he thought ironic. 'I couldn't use a NAAFI as a Bevin Boy before I was ill, or afterwards when I went back to the pit. But here I was, a civilian on an army site, and allowed in. Where was the sense?'

It was at the railway stations and on the trains that Bevin Boys most keenly felt a sense of ostracism, as Dan Duhig has never forgotten:

> There'd be hundreds of troops waiting for a train, with Toc H and the Sally Army there giving them a cup of tea and a sandwich. But not me. I remember being on a station – York I think, I changed there – in winter, a hut where you could go in for a warm, but not if you weren't in uniform. It was bitterly cold. I went to the end of the platform where there was a red paraffin lamp and tried to get warm on that. The trains were always full of squaddies, you in civvies. 'Hey, mate, why aren't you in uniform?' You can't keep explaining.

Adds Les Thomas:

> I was coming home to London from Yorkshire with some other Bevin Boys and there were women sitting close by in the carriage making remarks about young men still not being in the armed forces. This conversation was obviously meant for our hearing. We took no notice. Then an American air force officer joined us. He asked what the women's comments were all about. We explained why we weren't in uniform and the way the Bevin ballot worked. In earshot of the women he said something like 'I think you're doing a great job for the war effort and I'd rather have mine than yours.' We loved every minute of it.

Coming through a station barrier could be a problem. Les Wilson lost his identity card in Wales and got another, which was filled in with his lodgings address in Wales. On his first visit home to Brighton a policeman asked him for his ID, which he duly produced. Why a Brighton lad with a Wrexham address? And why in civvies? Fortunately a ticket collector at the station was a neighbour and vouched for him. 'What a knees-up,' he remarks. Owen Jones didn't even have his ID on him when he was stopped by a Military Policeman as he left King's Cross for Victoria, on his way to Sussex from Mansfield. 'I wasn't sufficiently savvy to carry it and I had nothing to show I was a Bevin Boy,' he explains.

'You a conscientious objector?' 'No, I'm not a conscientious objector.' But being the person I am I added that I admired conscientious objectors, saying I thought it took guts for them to do what they believed in. Naturally this caused deep suspicion with the Red Cap, whose cap was balanced by its peak on the bridge of his nose. 'Last three,' he demanded. This meant nothing to me but, I discovered, was how a soldier was identified – by the last three digits of his military number. I was finally let go, but it was an uncomfortable experience. Galling is a better word.

'I got that when I got back to Hastings on a pass,' adds Dan White '"You a conchie?" Or from my mates' girlfriends, "Why aren't you in uniform?", and those oh-yeah looks between themselves.' One of the less pleasant aspects of the war on the Home Front was that many people were all too ready to believe that any young man in civilian dress was a draft dodger or perhaps a deserter. More than a few Bevin Boys returned to their collieries upset that on leave they'd been approached by strangers and asked 'Why aren't you in uniform?' According to Stan Payne 'most of the questioners were women, almost certainly wives, mothers, sisters or girlfriends of serving men and were suffering under the stress of the times.'

It was hurtful to be in civvies and see old friends in uniform. Asked the familiar question by two former school chums, one in army uniform, the other navy, at a dance in Chapelhall, Lanarkshire, as he sat chatting to a couple of girls he'd once worked with, George Ralston 'stood up and whispered that my job was "hush hush".' Understanding nods. Say no more. Ralston still felt belittled by the encounter.

There were less trivial incidents. Cases are recorded of Bevin Boys being assaulted – one seriously, by a soldier in a cinema, as he stood for the National Anthem, the assailant under the impression that another young man not in uniform must be avoiding serving his King and was therefore in some way insulting him. One or two cases are recorded of Bevin Boys being handed a white feather, isolated reminders of a foolish practice prevalent in the First World War.[6]

Such occurrences led Bevin Boys to ask the Ministry of Labour for a badge that indicated who they were and had the support of the *Daily Mirror* – in the First World War a badge had been produced bearing the legend 'King and Country' for men who, for one reason or another, were legitimately not in uniform. Bevin wasn't to be persuaded and gave his reasons in a letter dated 22 March 1945 to Sir Edward Campbell, at the Ministry of Health, who'd forwarded a letter he'd received from a Bevin Boy in Durham the previous month:

I should explain that careful consideration has already been given to the possibility of issuing a distinctive badge to mining ballotees, but that it was found impossible to accede to this request.

Invidious distinctions would be created by the issue of such a badge in respect of one category of workers in one industry. A considerable number of other workers in the coalmining industry with a preference for the Services which they have not been allowed to exercise, as well as workers similarly placed in other industries, would have a justifiable grievance.

As the war entered the endgame what was really on Bevin Boys' minds was: when are we getting out? The newspapers wanted to know, too. Asked at a meeting in Norwich in February Bevin was irritated:

There's a lot of shouting about the Bevin Boys, but the miners have been mining for years and some people do not seem to realise it. Coal has to be got just as other things. It distressed me to have to send the Bevin Boys to get it, but I have had to send a lot of boys to do other dreadful things, much to my sorrow, and I am not going to be stampeded by threats or slogans thrown out by the press, left or right.

Victory in Europe came in May and with it a two-day national holiday. Some pits shut down, others worked through for a bonus, one of them Gresford in north Wales where Denys Owen was on the afternoon shift but 'escaped on the back of a lorry to Liverpool to join in the jollifications'.

A considerable number of Bevin Boys packed their suitcases and headed for home, convinced their service was over.

The May ballot, actually drawn on the day Germany surrendered, was cancelled; a good thing too, said a *Times* editorial, it had 'cut across the adventurousness and fine spirit of youths who, almost without exception, would have chosen to serve in the armed forces'. The Bevin Boys who'd disappeared were ordered back. Those caught in the April draw reported miserably to the training centres.

One of this last intake, Joe Geoghegan from New Brighton, the seaside town separated from Liverpool by the Mersey, went to Annfield Plain in County Durham. He'd just started his six-year study for the Catholic priesthood and anticipated being back in the seminary imminently – up to 1945 all Church students were exempt from call-up, provided they did a certain amount of farm work, and were again from 1946. Geoghegan fell through the gap and became a Bevin Boy, first in Durham and then on transfer in Lancashire. 'Perhaps,' he says wryly, 'I was naïve in thinking the authorities had made a mistake which they'd correct.'

On 17 May Bevin announced tentative release plans for the three armed forces: three-quarters of a million men out by the end of the year, the majority older men with time in (six years for many) but including, as a priority, 60,000 building trade workers to begin the reconstruction of the country, 1,000 miners, and teachers. He had nothing to say about the release of the Bevin Boys.

In the coming weeks many of the volunteers began to leave the pits, John Platts-Mills going from the Yorkshire Main to be elected as a Labour MP[7] in the General Election that turned Churchill out of office (and in which the overwhelming majority of conscript and optant Bevin Boys were too young to vote). Bevin became Foreign Secretary in the new Labour government and handed over his Ministry to George Isaacs. Japan surrendered in August. Another two-day public celebration, though a more muted affair than the one in May.

Details of the release plan for those not already singled out remained unsettled. When the demobilisation debate took place on 22 October 1945, nearly 2,000 Bevin Boys converged on the House of Commons while it was in progress. Little emerged from the debate other than that the three services were to be let go at different times because of their different sizes. The army, being the biggest, would take the longest – and its involvement in the occupied countries also had to be taken into account. Again no mention of the Bevin Boys – though Isaacs met a deputation from them.

The following day Flight Lieutenant William Teeling, who'd become Conservative MP for Brighton after war service and was one of the many MPs implored by Bevin Boys to get something done, made a lengthy and impassioned address to the House. There was no doubting his sincerity, but he painted what some of his listeners regarded as a distorted picture of the Bevin Boys' circumstances and he prompted acrimonious exchanges by claiming that they would be 'the first to be on the dole and also physically wrecked'. Billy Blyton, thirty-two years a miner and the new Labour MP for the Durham mining town of Houghton-le-Spring, was particularly put out. 'When I heard the honourable and gallant member talking about sandwiches and sweat, I recalled sweating without sandwiches,' he said. 'I remember that I coal-hewed five days a week and took 35 shillings home for a wife and two children as a result of the policy that was pursued in the coal industry by honourable members opposite when they were in government.'

What was clear from the second round of the debate was that Bevin Boys weren't going to be getting out soon. While there'd been a four-month surge in the workforce, principally men coming back from the now defunct munitions factories, natural wastage had again taken the total below 700,000. 'We cannot afford to lose a single man in the industry this winter,' declared Ness Edwards, the Parliamentary Secretary to the Ministry of Labour (and, coincidentally, another ex-miner), who dropped a bombshell into the laps of those Bevin Boys who'd been on their way to the RAF and navy before the ballot intervened. The government, he added, was minded to tie all Bevin Boy releases to the army's timetable.

Anger erupted. Hundreds of conscript Bevin Boys wrote to anybody they could think of. Surely, if they'd been destined for the navy or the air force, fairness dictated that their release dates should be the same? Many men who'd left the forces for the pits under the Bevin scheme were writing letters too: it emerged that time in the army and navy wouldn't count towards their release dates, but, inexplicably, would for those who'd come from the RAF.

At the end of November, George Isaacs confirmed the conscripts' fears: their release dates were to be tied to the army's, on an age and service basis. And with the army letting men go at the rate of 35,000 a fortnight, the earliest any of them could expect to be out was late summer 1946. Bevin Boys were in for the long haul. 'It was frustrating to see friends of mine called up for the navy like me released after two years, more than a year quicker than the army,' says Brian Folkes. 'The end of the war seemed a total irrelevance in what began to feel like imprisonment.' 'My younger brother went into the RAF after me and was out before me,' adds Harold Gibson. 'What can you say? The good thing about it was that in the winter of '47 he'd meet me on Blackburn station after my shift with a flask of coffee and a teacake and off we'd go to Ewood Park to stand on packed snow cheering on the Rovers.'

Absenteeism in the pits during 1946 was 14½ per cent among all workers (18 per cent among those on the face), but twice that among the Bevin Boys.

Prosecutions and fining for absenteeism had been some brake on the behaviour on those who wanted to step out of line, but these had gone with all the other wartime measures. Now everyone was fed up and discouraged by the chain of events, and even the committed like Alan Gregory began taking a day off, though when he did, 'which wasn't often, I told management I was going to. Nobody hollers too much if you do things right'. Probably the majority now missed a shift a week, finding that once tax was taken into account they earned nearly as much for a five-day week as a six, and a rebate some weeks later almost made up the money lost. But many missed more and even refused to do night shifts. 'What you could call "subsistence mining" became the new pattern of work,' says Geoffrey Mockford. 'You earned just enough to keep going – which was about three shifts.'

Some collieries threw up their hands. Others – mostly the big ones – still tried to maintain discipline, though their options were limited. Phil Yates' absenteeism at the Prince of Wales in Pontefract got him sent off as a punishment to Ackton Hall colliery in Featherstone.

> I came back one weekend to find a notice on my lamp, 'No longer required because of absenteeism, reason for transfer'. It wasn't much of a punishment. I was still in the same hostel, Hightown in Castleford, and Ackton was so chaotic a lot of Bevin Boys were hardly there. We didn't just take weekends, we took long weekends – miss the Friday shift and come back on the Tuesday. The miners weren't the least put out. 'Good luck to you,' they said. The pit didn't do anything about it.

Manny Shinwell, who'd taken over from Lloyd George as Minister of Fuel, tried everything he knew to bring absenteeism down. He tried being tough and he tried being accommodating. In April he threatened miners and Bevin Boys with sanctions, adding that he believed midweek sports fixtures were a major cause of absenteeism, and should be cancelled in the interests of getting the coal to keep industry humming.[8] Then in August, when Arthur Horner, general secretary of what was now the National Union of Mineworkers, said more men were needed (so many miners had retired, natural wastage had jumped to 40,000) he announced that he'd get 30,000 ex-pitmen out of the services ahead of their turn if they agreed to come back to the industry for at least six months. When the uptake was negligible and the NUM was saying that the only way to attract newcomers was to give mining 'a new status', he grew exasperated and told them there were enough men if only they'd work – but then started recruiting unemployed Irish and Poles to bulk up the numbers.

The Conservative MP for Aberdeen, Bob Boothby, declared that the government had no idea how to get the best out of the miners and proposed they 'be treated as a corps d'elite and given houses, higher wages and more food than any other class of manual worker' – ignoring that the miners still enjoyed the rationing privileges they'd enjoyed during the war and that when bread had been rationed in July 1946,[9] they'd been issued with extra coupons (which, according to John Wiffen, 'always gave us more bread than we ever needed'). Boothby, a notorious self-publicist, was, in fact, only anticipating the government announcement that shortly followed: that shops in mining areas would receive extra sugar and fats for making cakes, more supplies of tinned meat and fish, and the weekly meat ration for underground miners would increase by 75 per cent, from 1s 4d to 2s 4d.[10] Miners' families would also get priority for housing.

With considerable lack of discretion, Arthur Horner, general secretary of the NUM, greeted the news by saying miners 'must have more food and more unrationed goods, even if some other people of much less importance have to have less'.

The government announcement was made in October, when as part of the attempt to win over the miners Shinwell had written to them – and the Bevin Boys – pleading for absenteeism to stop. Once again the situation was reaching critical levels: coal stocks were 5½ million tons lower than a year previously – indeed below any year for which there were recorded figures – gas and electricity stations held less than three weeks' stocks, railways, coke and steel works less than two, and the Lancashire cotton mills were among the rising number of businesses intermittently closing because of a lack of coal.

The government was looking for 200 million tons in the year, even 220. It got 194.

'If I had a supply of wireless sets, pianos, motor bicycles and other things to sell to the men on hire purchase I could reduce absenteeism,' a colliery manager was quoted as saying, though it seemed that whatever inducements the miners were given they were never enough.

The only incentive the Bevin Boys wanted was their release – and that receded from them with another government announcement: because of commitments in the Far East and Arabia, as well as the Arab-Jewish conflict in Palestine and pre-independence civil war in India, many soldiers were being held back and, in consequence, only three army groups would be released in the first six months of 1947.

Around 1,000 Bevin Boys converged on Wolverhampton in November to discuss this turn of events. A mass strike was mooted. The majority were reluctant, though subsequently a few hundred in Cannock Chase downed tools for two days. Early in December two or three conscript Bevin Boys at Bentley colliery in Yorkshire, one of them Morry Pearce, 'had a little powwow in the pit canteen' and decided to form a Bevin Boy association to lobby for release.

There'd been two attempts to form an association in 1944, though the object of these was purely welfare. The first, the idea of a Nottingham businessman, foundered through lack of interest; the second, that of a scoutmaster in Wallasey (a town on the mouth of the Mersey), because it incurred the wrath of the miners' Federation whose president, Will Lawther, wrote to Lloyd George suggesting he should take immediate steps, 'otherwise, I am afraid, our members will do so'.

This time, Pearce wanted the miners on side and persuaded Ernest Jones, the area secretary of the NUM, that 'this wasn't a breakaway union, we weren't picking a fight with the miners but with the government'. Jones was sympathetic, telling the *Yorkshire Post*: 'If the Bevin Boys are simply asking for the same treatment on demobilisation as men in the Forces, the union will give them every help.' On his instigation the Bevin Boys were allowed to erect a table outside the Bentley gates and canvas for members, ex-forces excluded – 'they were in the pits because they'd chosen to be,' Pearce explains. 'They could fight their own corner.'

Three days later a union official at the colliery, 'the very aggressive Mr George Hukerby', persuaded the branch that a breakaway union was exactly what the Bevin Boys were after, the miners threatened to strike, and Jones changed his position, telling the *Daily Express*: 'We cannot tolerate anyone organising miners other than in our own union. The union is the body to deal with any grievances by miners, whether they are Bevin Boys or not.'

The association spread to others pits in Yorkshire and then in fragmented fashion to other coalfields; the miners stayed their hand. Action committees were formed in some of the hostels. More lobbying of MPs. Correspondence with the Minister of Labour – which Isaacs didn't even acknowledge. Nothing changed.

At Christmas absenteeism of miners and Bevin Boys was through the roof. In some Midland collieries on the day after the break it was as much as 80 per cent (but attendance to draw pay was 95 per cent); out of 200-odd workers on the day shift at Clifton in Nottingham, only eighty men turned in, and only fifty on the afternoon and night shifts. In Shropshire attendance was 50 per cent, South

Staffordshire 40–60 per cent. At one of the largest collieries in south Wales only a third of the 6,500 men reported. During the week the country's coal production more than halved.

* * *

Psychologically the end of the war and the indeterminate length of time they'd be kept in the pits resulted in a proportion of Bevin Boys resorting to more than absenteeism.

Many in the hostels took a day off and hung about, playing desultory games of table tennis and billiards. For some, a day became another and another; apathy set in and they rarely turned in for a shift. Geoffrey Mockford remembers one, who was known as 'the Porridge Boy' because he ate porridge in prodigious quantities, often filled with crusts. But numbers of Bevin Boys who didn't work regularly and in consequence couldn't afford meal tickets lived on a diet of porridge, bread and tea, which were free. Mockford also remembers

> a public schoolboy liked by everyone but the most unsuitable person to send down a mine. He didn't go to work all the while I knew him, about a year before I was demobbed. He was a botanist and was going to Cambridge after his release. His time at Abbott Road followed a predictable cycle. As he didn't work he couldn't pay the hostel charges and had to see the manager every Friday. After four weeks of non-payment he was refused further credit and told to leave. To raise the money for his train fare he painted small pictures of birds, which he sold for half a crown. I bought one. When he had enough he left and went home. His father then paid the bill and a few days later he'd return. And the cycle started all over again. As far as I know no action was ever taken against him by the authorities. He was still there when I left.

The hostels couldn't usually be that lenient. There was a stream of permanent evictions. Those who were turned out presumably drifted away.

Some Bevin Boys who stayed and did limited shifts took part-time jobs. Flight Lieutenant Teeling heard of one Bevin Boy who drove a taxi when he should have been on shift. It was well known in Gedlington colliery in Nottingham that an ex-forces Bevin Boy with two butcher's shops in the town came and went as he pleased, the manager, provided with regular pieces of meat, turning a blind eye. According to his obituary (in August 2001) Paul Hamlyn – the publishing multimillionaire and Labour Party benefactor – a Bevin Boy in south Wales, moonlighted as an excursion coach driver, a weekend rep for a paperback company, and as a freelance reporter (weddings and football matches) on the *South Wales Argus*. More commonly, Bevin Boys went digging potatoes on local farms or fruit picking. 'We had around 60 Bevin Boys at Cossall [in Nottinghamshire] and a lot of them were forever swanning off,' says Alan Gregory. 'Half the buggers were never there in the fruit-picking season.' In the summer of '46 when the pits shut for the annual break, Ron Bown and a fellow Bevin Boy from Hilton Main colliery near Wolverhampton went pea picking at Shenstone near Lichfield under the national 'Lend a hand on the land' scheme, but instead of staying for the week, stayed two. 'A paid holiday – a change for a Bevin Boy,' he remarks.

In mid-1946 another of Shinwell's surveys revealed than 8,000 Bevin Boys hadn't been to work in a month, 'and some are engaged on other business'. As he'd threatened, he subsequently took action, shipping 'the undesirable elements' out to the army, with time in the pits (over two years in cases) not counting against demob. The national service of such Bevin Boys, one way and the other, became about five years.

Staying and working in other jobs was a middle course that some other Bevin Boys spurned: they just refused to serve any longer. One was Ernest Noble, though it took being badly injured in a rock fall down Frickley pit in Swillington, Yorkshire, to make up his mind. In the accident, he sustained cracked ribs and muscle damage to his left arm. Unfit for six months, he returned to his job in the railway office in Bradford, subject to remedical. When he was pronounced A1 and redirected to Frickley, he said no, supported by the Railway Clerks' Association, and was sent into the army, where he volunteered for the Paras – 'truly a conscription service of upstairs, downstairs'. Other Bevin Boys, like Warwick Taylor simply decided enough was enough. He went absent without leave from Oakdale in south Wales in February 1946 and waited for the knock on the door. 'I'd loathed being in the pit and wasn't prepared to go on,' he admits.

> Usually it took a few weeks before the law caught up. I was only at home a few days when a policeman arrived and ordered me to report to my nearest labour exchange. They said I'd got to go back. I said I wouldn't. 'It'll mean a tribunal and imprisonment.' 'I don't care. I am not refusing to serve, but I wanted the RAF and that's what I still want.' Things were beginning to relax. They let me go.

But not without a fright: 'I got papers for the army, a camp in Canterbury. I sent them back, sorry, you made a mistake.' The following month Taylor was in the RAF and trained as a radio operator, 'which turned out to be a redundant trade with the war over, so I had to remuster as either an RAF "Snowdrop" [military policeman], cook or driver. It wasn't much of a choice – but being a driver at Lytham St Annes was a lot better than being a Bevin Boy.'

Ken Sadler was another who felt unable to continue. At ballot he'd tried desperately to avoid being directed to the pits by attempting to enlist in everything from the colonial police to the merchant navy. Now, like Taylor, he felt he'd reached the end of the road at Blagdon colliery in Durham.

Two friends had gone out, one on compassionate grounds, the other because of persistent absenteeism and he was now in the army in Germany. His way seemed to be the solution. I walked in to see the colliery manager and told him I was finished. I suppose his jaw dropped. I was cocky, and very rude. I told him what he could do with his coalmine and said over my shoulder as I left that he knew what he could do with me – send me off to the army. Around that time, I believe, there were 5,000 deserters from the pits in the North East alone.

I was very naïve. I went down the labour exchange to sign on, would you believe. When they heard I was a Bevin Boy they were puzzled what they could do. They couldn't give me a job and they couldn't give me any money. I was living at home so I lived off my father who was retired – he'd been a tram conductor most of his life. Then I got a Ministry of Labour letter to report to a tribunal in Gosforth outside Newcastle: a regional colliery court made up of the mining community. It's always puzzled me why mining was the only industry that had a private court.

My brother was a sergeant in the Signals and happened to be home on leave. He tried to speak for me at the tribunal – he'd had a word with his commanding officer who was prepared to take me – but they told him in no uncertain terms that he had no authority in their court. A few days later I received an order to return to Blagdon, warning me I faced imprisonment if I didn't. I was so incensed I wasn't going to obey the order. My brother convinced me I had no option.

A few months later Sadler, a spectacles wearer, tried again to escape by claiming his eyesight was deteriorating. 'I was sent to an optician for a test and eventually issued with a medical card marked across it in red "Fit for coalmining." I'd failed again.'

At all stages, right from the appeals tribunals, through training and into the pits, a proportion of Bevin Boys tried everything they knew to get out. Some failed; others succeeded. A friend of Jim Bates in Hem Heath, Stoke, was so determined to get into the RAF that

one day he downed tools and hid in a manhole near the pit bottom. The end of the shift came, his lamp check remained unclaimed, so search parties were sent out and every road in the pit scrutinised. Coming back to the pit bottom a miner shone his lamp in the manhole and there, under a piece of sheeting, lay my good friend fast asleep. He got into the air force, jet-propelled!

According to Peter Allen, 'one particular character from London' at Ryhope near Sunderland, who most certainly wasn't what we now euphemistically call gay, accosted the son of the canteen manager on a number of occasions and twice chased him around the tables late at night. He obtained his discharge. Incidentally, so did another Bevin Boy who was clearly bent. He outraged the miners by getting out his compact and lipstick during snap. He was also packed off before he came to any harm.

But exaggerating, or inventing, a medical condition was the most popular try-on and it increased markedly with the end of the war. Ernie Jefferies took that route out of the Glycorrwg pit in south Wales.

I worked bloody hard for 18 months, but the war was done, finished. And so was I. Someone said the way to work your ticket was to go to the doctor with a bad back. So I did: 'I have a bad back.' I was sent to a medical board, say if it hurt where they touched you in certain places. I dutifully groaned and moaned. And I was out.

Peter Rainbow chose a more extreme method to get away. 'A good worker and a good timekeeper' as his transfer papers from Leycett colliery near Stoke to Holditch colliery between Stoke and Newcastle-under-Lyme made clear, for a number of months he earned well as a collier.

But eventually I asked myself where was I going? It was all hard graft and no future. Obviously they weren't going to let a fit miner quit – so I invented my own medical disability. I obtained a small bottle of concentrated nitric acid from a local chemist and back in my lodgings carefully applied it to my ankles. I repeated the treatment. Within ten days my skin was raw, like a typical case of dermatitis. The medical officer had no hesitation in issuing me with a discharge certificate. Needless to say my 'dermatitis' cleared up in a few weeks, by which time I was back in the West Country. Nobody raised the issue of the army, and I certainly didn't.[11]

I wasn't ashamed of what I'd done, not in the least. I'd done my bit.

Doctors sometimes of their own volition helped a Bevin Boy to the 'exit', as Reg Fisher was astonished to discover, two months after the war ended and only nine months into his time at Ramcroft colliery in Derbyshire.

There was a terrible fall in the pit and I was lucky not to be under it – managed to scramble into a safety hole as the lot came down with a wallop. I was really shook up. I went to the GP I'd popped into before with a touch of bronchitis and catarrh, nothing much, and he'd given me the odd day off. Now I asked him for a chitty for a couple, to get over it. He said, 'I think we can do better than that. Other lads come to me with different things, but you've always come with the same thing', and he gave me a letter: *Unfit for the mines*.

Jubilant, Fisher went home, was examined by of his medical board in north-west London, was called back in and told they couldn't find anything wrong with him; but on the strength of the doctor's letter they graded him B2 and dismissed him from the pits.

> I tried to look disappointed, how sad, but jumping up and down inside. 'Thank you very much,' I said, thinking I was off. But then they said, 'Go upstairs for your interview.' 'I thought this was the interview.' 'No,' they said, 'this is the outcome of the medical, your interview is in Room 2 upstairs.' Having been there before, I knew what that was. The RAF. So I was a free man for a flight of stairs. The RAF wasn't so bad. At least I was above ground.

In south Wales, Deryck Selby encountered a GP of the same disposition as Fisher's. Having badly strained his wrist in the Empire pit after three or four months' service, he anticipated that the doctor would sign him off perhaps for a week. 'Instead, the gracious man said, "You can't work in the pit with that," and wrote: *Unfit for present occupation*. I didn't argue with him. I went to the labour exchange next morning, got a travel warrant and straight home.'

Whether a Bevin Boy's injury or illness was real or not, if he was subsequently passed A1 by a medical board he should have been returned to his colliery. There was, however, considerable inconsistency in the system and at least some boards were inclined – like some magistrates who tried Bevin Boy refusniks – to let a man who pleaded for the forces have his way, or took the decision to send him to the army, anyway, particularly after the war when it was both overstretched and losing men hand over fist through demob.

Some men, who in other circumstances might have been discharged altogether, didn't escape the army, including Dennis Faulkner whose intestinal problems had him dismissed from the Deep Duffryn colliery in south Wales. He was so certain that he was a civilian again that he applied for his BBC job back. On VE Day he attended a church service of thanksgiving which, he remarks, 'for me turned out to be a sort of double-edged sword – they said I was fit for the army and put me in the Royal Signals for the next 2½ years.' Other men were happy to get into uniform. Deryck Selby was posted into the Royal Artillery. Ernie Jefferies went to the Queen's Royal Regiment. 'At last,' I thought, 'my chance to see abroad, do heroic things. And I finished up at an officers' training unit teaching fieldcraft in Ashdown Forest. Life's daft, isn't it?' Ken da Costa, out of the Arkwright pit near Chesterfield for six months with a face poisoned in an underground accident, was another who went to the Signals, after he was better – even though a Harley

Street specialist said he was fit enough to go back to the pit. 'Another mad thing,' he adds. 'I was down to go to Burma but the night before we had a medical and the MO noticed my toes were curled under. Off the draft, B2 – which would have had me turned down for mining in the first place!'

There was the same rush home for VJ Day as there'd been three months earlier for VE Day; the difference was, this time, a proportion of Bevin Boys didn't come back.

During 1944, with the war still on, most deserters were rounded up, though not always quickly: one Bevin Boy ran away from Oakdale in south Wales within weeks and went to London 'doing blitz work' for a year before the authorities found him. Or rounded up at all: as long as a Bevin Boy found ways of surviving without an employment card, he had few obstacles. He had his identity card and certificate of national service registration; there was nothing to single him out to a casually enquiring policeman. Doug Ayres encountered a one-time fellow worker, who'd disappeared from the Homer and Sutherland colliery in Stoke, working as a bus conductor in London three years later – Ayres, still a miner, was on leave. 'We had a Scotch Bevin lad in Durham got a local girl pregnant and did a runner,' remembers Bill Hitch.

> He got into the army and was sent to Egypt. How he did it I don't know. They collared him in the desert after about six months – he was back in the pit before the baby was born. He got married and lived in the same street as me when I was married.

Albert Mitchell wasn't rounded up after he did a runner from the Lady Windsor colliery in south Wales – he turned himself in.

> Only been there six or eight weeks and broke my foot in a roof fall. Into Caerphilly hospital. Got out, thought sod this and went home. I wanted to fight for my country, not dig for it, simple as that. So I went home to Bristol. The police came round for six or seven months but I was off, part-time work on farms picking apples, anything I could get. In the end I gave myself up. I was courting a girl at the time and she didn't like me being a fugitive. Did a month in Horfield prison in Bristol. Draft dodgers and all sorts in there.

Thereafter it was the Service Corps, before a transfer to the Catering Corps, with service in Egypt, where Mitchell 'met many miners who'd have willingly gone back to the mines but weren't allowed to'.

It was 1946 when the number of desertions peaked, by which time men had put in eighteen months or two years. It was then that John Brinsmead, a printer's apprentice from Worthing, walked away from Hirwaun in south Wales. 'I'd gone to work and I'd worked hard,' he says.

I took time off like everyone else, but we didn't push it – usually when we had a hangover. But the war was over and I just thought 'I'm not going to put up with this for £3 10s.' Several of the lads had been in prison, so I thought they could put me inside too. I just came home.

Not long after, possibly days, I got another set of call-up papers telling me to report to the Rifle Brigade in Winchester – some kind of cock-up, I imagine, but I wasn't looking a gift horse in the mouth. That's what I did – and spent another 18 months in occupied Germany.

David Roland called it quits about the same time, after his friend George survived a rock fall in the South Normanton pit in Derbyshire.

We got him to the surface and found he was covered in bruises, that's all. I'd expected every bone in his body to be broken. Absolutely miraculous. He came out shaking. 'I've had enough,' he said. I was scared by what had happened to him. 'Me too,' I said. We went home to London.

I have an idea George went back to the pit and married a miner's daughter. Me, I was out of authority's grasp for five or six months, working as a general labourer and glazier for a building company. Then I fell off a ladder and ended up in hospital, where they took my details. Next thing, the boys in blue were round.

The question of returning me to the pit never came up. I was packed off for training to the Military Police, which was bloody awful. I was small for a Military Policeman and I looked different – my looks are Mediterranean, anywhere I go on holiday, Italy, Spain, wherever, holidaymakers ask me for directions. So I got into a lot of fights I could have done without. I transferred into REME.

Dan White got fed up as more and more Bevin Boys disappeared from the Ocean colliery in south Wales and finally decided to call it a day. 'There were about 70 of us to start with but by September '46 not many of us were left,' he says.

One day I thought, 'That's it. I said to the Bevin Boy who shared the lodgings with me, 'You know my address?' 'Yes,' he said. 'Well you forget it,' I said. 'I'm off.'

I went back to Hastings and didn't have a regular job for months – as soon as they asked for my cards I was sacked. Building sites, plastering – I did ceilings in pubs, then became a stoker at Ore power station for a couple of days but sod that, it was worse than the pits: trucks used to come alongside of me and dump the coal, most of it dust, to shovel into the oven. In the open, freezing on one side, boiling on the other when the doors were open. Stuff it.

I was helping a bloke shift furniture to Brighton and we had a meal in the town. I was walking back up the main drag and I saw a hoarding, recruiting for the police. Walked in – all the blokes there were Welsh. What have you been doing? Working in Wales. They assumed I was Welsh, very friendly, so as far as they were concerned I was Welsh. Interview, passed, railway warrant to Cannock in Staffs to the training place, an ex-prisoner of war camp, and down to Lancing as a copper.

My mother got letters from the Ministry of Labour that I was avoiding national service – all that sort of cobblers. The police went round her house a number of times and she lied she hadn't seen me. Then after three or four months she told them I was in Lancing and I got hauled up in front of the West Sussex chief constable in Chichester. 'I hear you've been avoiding service in the pits,' he said. 'That's one way of putting it,' I said. 'Wasn't much of a job, was it?' he said. 'You can say that again,' I said. And then he said, 'You won't hear any more about this.' And I didn't.

Many Bevin Boys agree with Dan White that most of their fellows disappeared; many more, however, especially in the English pits, say that the majority stuck it out until their release dates set them free. According to Arthur Horner, when the miners' leader in Wales, by the end of 1946 desertions were a daily occurrence and in some pits in south Wales, Durham and Scotland, virtually all the Bevin Boys had gone. 'Of the 45,000 boys who were directed to the mines,' Horner said, 'nearly half of them have thrown up their jobs. They are being sought in every part of the country.' Newspapers latched on to his statement and claimed that Bevin Boys who'd deserted were engaged in the black market. Two Bevin Boys, 200, 2,000? And Del Boy-ing in what, coal? No substantiation was forthcoming. There seemed to be no substantiation of Horner's figure either.

Nine

The End of the Beginning

The 'new morality' that miners' leader Arthur Horner had begged his membership to adopt wasn't forthcoming in 1945 or 1946. The hope, as 1947 approached, was that nationalising the mines would bring a new beginning.

The handover from private to state ownership occurred on the first day of the new year. At 11.30a.m. at the tidied-up pit tops (some of which, albeit briefly, even sported flowers), the hooters sounded, union banners were unfurled, bands played, speeches made and plaques unveiled, proclaiming: 'This colliery is now managed by the National Coal Board on behalf of the people.' The sense of expectation among the miners, Bevin Boys remember, was palpable. 'We're the bosses now,' they said, and joked about going down the pits wearing bowler hats instead of safety helmets.

The NCB made the symbolic occasion of nationalisation day the time to replace lighthouse lamps at collieries still using them, including Stan Payne's at Welbeck in Nottinghamshire – although here the plan had to be brought forward a few days:

It just so happened a local lad suffered a crushed foot – while coupling tubs his old lamp gave up, he missed the coupling hook and in total darkness the tubs ran backwards over his foot. Next day, protest meeting in the yard. Management said they were powerless to override orders from celestial heights as it were. Union demanded immediate issue of cap lamps on threat of strike. Management unbending. Show of hands. Instantaneous strike.

With the summer shutdown five month ahead it would have been awkward to save enough to get home, so my friend Sid and I fixed up to work on Gledthorpe farm just outside the village. We worked there next day for 12 hours and received 10 shillings each. That evening word circulated that the new cap lamps would be issued next day and work resumed.

On the day itself at Bolsover colliery in Derbyshire, Victor Simons remembers

A group of Bevin Boys had just come off day shift and were waiting at the bus stop in biting cold when the bigwigs who'd help raise the NCB flag set off in their cars for somewhere else and sailed by. My mates weren't amused – a lift would have been a nice gesture on such an auspicious day – and they made it clear in no uncertain terms in a letter to the *Derbyshire Times*. It didn't go down well with the powers that be.

Geoff Darby retains a more colourful memory from Lepton Edge colliery in Yorkshire: of 'the eccentric George Elliot climbing the headgear and pulling down the NCB flag – proclaiming he hadn't yet been paid for his pit!'[1]

The weather had been kind to Britain in the first three winters of the war; the exceptionally mild one of 1942–43 even helped to conserve coal stocks. But the next was severe and the following two worse, with severe snow storms blotting out parts of the country. Winter '47 came in bitterly cold – and right after nationalisation day turned into the worst of the century. From mid-January to mid-March it stayed below freezing almost without break and the snow and ice caused total paralysis. Whole areas were under 3ft of snow and cut of by drifts of up to 20ft; food and milk supplies struggled to get through and the RAF made food drops. Steam trains trapped deep in snow blocked main lines for weeks on end. Thousands of miles of road were impassable, under up to 4ft of ice that the government tried unsuccessfully to shift using flame-throwers and even jet engines. Gales kept ships bringing coal from the North East to the South in port. Restrictions were imposed on electricity, power cuts were frequent and gas pressure reduced. Factories and businesses shut down and millions were out of work. Cigarette manufacturers ceased production; beer manufacturers too – parts of England would be dry for a month when the thaw came. The BBC reduced programmes, newspapers were cut back to wartime sizes and league football was all but scrapped (when it resumed it went on until June).

The country needed coal – but in February 300,000 tons were stuck in railway sidings, the wagons fully buried in drifts. The stock of wagons had been depleted during the war and was now slowly being replaced, but 200,000 old ones were in workshops and those that had to be scrapped outstripped repair and replacement. To help turn round the wagons that were still running, the rail unions put virtually all its workforce on the job – suspending 'out of grade' working for the first time since Dunkirk. The shortage, however, was critical and the majority of collieries couldn't always be kept supplied. Without wagons all they could do was stack the coal they dug at the pithead.

While the unfortunate NCB and the government struggled with distribution, the miners were urged to step up output. In most collieries they responded, where they could – some collieries were snowbound and out of commission for much of the time, travelling conditions were so bad that at others too few men sometimes arrived to make working a shift possible, while yet at others where there were enough men there were no wagons and no more stacking space at the pithead. Nonetheless, there were record attendances on Saturdays as well as on weekdays and many men worked on Sundays, too. It was on a Sunday – 9 February – that Britain came within forty-eight hours of the failure of the national grid, but the disaster was warded off by the miners' efforts (and, it should be said, those of the railway workers, coal trimmers and seamen). In the eleven weeks to 22 February, despite being hampered by regular power cuts, miners dug 39 million tons, more than 3 million above the same period in 1946.

Like the miners, the Bevin Boys rose to the challenge, like them struggling to get to work – and struggling, sometimes, even to get in and out the door. There was a big dip between the Easington hostel in Durham and Bevin Boys waded through the snow to reach the main road. 'Once we had to shovel our way out to go on shift,' says Derrick Warren. 'And the lads returning from the earlier shift had to shovel their way back in – that's how hard it was snowing.' Those who made their way by public transport were grateful for the dedication of transport staff who frequently worked miracles. Reg Taylor remembers the driver of his double decker bus who always turned up on time near his home in Bradford, even in a blizzard. Only once did it appear to Taylor (sitting with his lemonade bottle of hot tea between his thighs) that the conditions would defeat him.

We were confronted by a 15-foot snow drift across the road between the gable ends of two houses. It seemed obvious we'd have to turn back. But the bus was powered by a big diesel engine that stuck out a foot further than the original one, and the driver was of the stuff of heroes. He got out, had a look, decided the drift was thinner at one end and he could make it – which he did, with a lot of grinding noises and mighty shudderings.

At the Nutters colliery few believed the story – 'still less when on the return journey, as I was saying "You'll see in a minute" for the umpteenth time, I discovered the snowploughs had removed all traces of my snow wall'.

Bevin Boys remember the stark beauty of the pitheads that winter, the great winding wheels emphasised by their blackness against the whiteness of the landscape, the ever-present coal dust sprinkling on every fresh snowfall. But above all, they remember how glad they were to get down the pit (not all did: some unfortunates were detailed to free tub wheels frozen to the rails by using blowtorches). Going to the cages passed the braziers that ringed the mouth of the pit and burnt day and night to heat the freezing air drawn down the shaft made Norman Brickell at Monckton in Yorkshire think 'It was rather like entering the fires of hell'.

Glad as they were to get down, those working at the pit bottom were less lucky than those who went inbye, where there was warmth. If the braziers were doing their job, the ice melting on the shaft walls fell like heavy rain, making for a miserable shift despite oilskins. If the braziers were less efficient the pit bottom felt almost as cold as the surface, however many layers of clothing men wore. It was so cold down the Shipley Woodside pit in Derbyshire, says John Potts, 'that the cage onsetters had icicles on their eyebrows and the end of their noses'.

The pit bottom could be particularly hazardous in the winter of '47: periodically, one of the huge icicles, several feet long and inches thick, that formed on the shaft walls, broke away and came crashing down. Bevin Boys pushing tubs in and out of the cages kept an ear cocked. 'You could hear the snap and the swoosh of them coming and you moved – fast,' says Potts. Icicles shattered on the tops of the ascending and descending cages, showering the occupants with fragments; in Durham, an icicle killed a collier riding the cage.

The work at pit bottom was even harder than usual: the ice and snow coming down on the tubs made the flats treacherous and boots had little traction – and the tubs were resistant to being moved at all, the lubricating grease on the wheels frozen like everything else. But at least the cold on the surface wasn't a shock to the system at the end of a shift, as it was for those who worked in the heat of the coalface. Frequently there were nine degrees of frost. By the time men at collieries without pithead baths boarded their buses, the sweat on their pit clothes had turned to ice and cracked as they moved. Doug Ayres, who did have pit baths at Bolsover in Derbyshire, remembers coming away with wet hair 'and as I walked up to the main road the hair froze on my head'. David Reekie loved his clogs, but cursed them in the snow as he came and went to Horse Riggs colliery in Yorkshire: 'The stuff built up between the two irons on the soles. Within a few yards it was impacted and rose higher and higher with each step. It was like walking on platform shoes. You were banging them off on every wall and lamppost.'

Minster of Fuel Manny Shinwell rushed from colliery to colliery urging men on. He came to Lepton Edge near Huddersfield, which was anxious, as Geoff Darby relates, to put on a good show:

> We had full sets of loaded tubs by the face, all waiting to go down an incline to impress the Minister. As the Bevin Boys started them off they shouted in unison, 'Coal for Shinwell.' Unfortunately, the front half and the back half of the first set weren't linked and the front half had no wooden 'sprags' pushed in the wheels to slow the batch down. Off they went and off the rails they came, tubs at all angles, pit props down, coal all over the pit. And Bevin Boys shouting in unison: 'No coal for Shinwell.'

Before the war miners could turn in for work, go down the pit and then be told to go home – without pay. Bevin had stopped that. If a man came to work and changed ready for it and there was none, he was paid his basic day rate – he had what quickly became called a 'Bevin'. Many miners and Bevin Boys had 'Bevins' in the winter of '47 – Geoffrey Mockford at Pleasley in Derbyshire had three together when there were no wagons. Six miles away at Bolsover, Doug Ayres never had one. 'The main railway line was immediately adjacent to our colliery sidings,' he explains. 'As soon as that was cleared, wagons appeared.' At Welbeck in Yorkshire Stan Payne turned in for a morning shift – leaving his co-lodger Sid, who'd decided to take an unofficial day off, in bed – to be greeted at the colliery gates by a notice chalked on a blackboard: PIT BEVIN TODAY. Payne quickly changed into his pit clothes (to meet the requirement of a 'Bevin'), reported to his deputy, then

> straight back up and dashed back to wake sleeping Sid. 'It's a "Bevin", mate. You've just got time to sign in.' Sid up and off like a flash, reports to his deputy. 'Ah, Sid, good lad, we need a pony driver on road maintenance.' Hard to get much out of Sid for several days.

Absenteeism among Bevin Boys was very low during the early months of 1947. Lodgings and hostels were so cold going to work was a better option. The heating in the hostel Nissen huts – a single hot water pipe running around the curve of the metal ceiling – was inadequate at the best of times but was now non-existent. Bevin Boys, with all the clothing they possessed piled on their beds, woke up to thick frost on the metal surfaces above them. Power cuts sometimes meant there were no cooking facilities. At Easington hostel, says Derrick Warren,

> Four or five of us volunteered to help in the kitchen and with candles and gas rings managed to provide toast, bacon and tea for everyone coming off night shift. The girls in the kitchen were so grateful. I didn't need to buy meal tickets after that. And I got my shirts ironed free.
>
> Someone from the Ministry came up in his chauffeur-driven car and said conditions were okay as far as he was concerned. We made it clear they weren't – it was bloody freezing, there was ice on the pipes. Everyone was standing on tables watching him have an evening meal, brought in by Bevin Boys. I don't know what went into his dinner that shouldn't: I was one of those standing on the tables. But it did: I know salt was used on his sweet instead of sugar. But he was a brave man, he ate the lot. How things went for him going back to London I don't know – some of the lads had put sugar in his petrol tank.
>
> Still, he must have been a decent man as well as a brave one. Next day paraffin heaters appeared.

In what was a state of emergency, the government temporarily stopped coal exports – which were so vital to the economy – and ordered supplies from the USA and Poland. And it urged everyone to limit their use of gas and electricity.

Industry complied. Householders refused. Exactly five years earlier (January 1942) the same appeal had been made and loyally answered, as was the appeal to burn as little coal as possible. Now the British had had enough. And they blamed the miners for what was happening. In turn, the miners' leaders blamed the extra trains that had been laid on at Christmas time which, they said, had hindered coal being transported; delays in the arrival of new mining equipment; a lack of strict supervision of the use of electricity once the emergency began; and, as ever, a shortage of manpower. The public, sympathetic to the miners' circumstances throughout the war and its aftermath and surprisingly tolerant of the strikes and absenteeism that blighted the mining industry, were having none of it. The stark fact was that the miners' behaviour in 1946 had meant that, for the second year running, the country entered the winter with a deficiency of over 5 million tons of fuel in the power stations and factories.

Coal hadn't been officially rationed during the war[2] and it wasn't now – but it was almost unobtainable; when merchants did have deliveries, people queued in the snow for hours to collect a 28lb bag. During the war there'd been some largely unvoiced resentment of the miners' traditional 'free' coal allowance; quite naturally those who queued and foraged for something to put in the grate were envious of miner neighbours with full bunkers or coal sheds. Now, however, for the first time, there was deep anger and calls for the allowance to be cut or withdrawn.

Those Bevin Boys allowed free coal by their collieries and able to have it put to good use took it without compunction (one Bevin Boy in Yorkshire bought half a hundredweight and dragged it on the bus to his girlfriend's, ignoring the covetous looks of the other passengers). A number of collieries that previously hadn't offered Bevin Boys the concession now did so, including the Nutters, where Reg Taylor paid for his monthly ton at pithead prices.

> One day I tramped home through the snow to see a horse-drawn cart dump a load of glistening black diamonds at our front gate for all the world to see. Some of the neighbours hadn't acknowledged me, much less spoken to me, for the years I'd been going back and forth in my pit dirt – there was a silly social stigma attached to me, I suppose. But my popularity soared. I was complimented on my work of national importance, and was there any chance of a bucket of coal?

Unknown to Brian Folkes, his father, a wine importer in the City, lost patience with the slow release of Bevin Boys and

> wrote to the King, complaining that my banishment to the far corner of the Empire, Durham, was seriously impeding my studies and my career – or words to that effect. With almost the speed of light I found myself transferred

to Betteshanger in Kent. Whether I was 300 miles from my home in London or 75 miles, which I now was, made little difference to my studies. But it was a change of scenery – and I did cycle home a few weekends.

The weather took a turn for the better as spring approached, but at the price of flooding and landslides. The wheat fields had been replanted; the floods washed the seeds away and with them any hope of a swift end to bread rationing, which was to remain until July 1948.

The spring brought no relief to the coal crisis.

An argument that batted back and forth throughout the war was whether or not the pits would be more productive if they worked five longer days and dispensed with the Saturday shift, which was woefully inefficient because so many men didn't show up. Now, a five-day week made sense to everybody, including the National Union of Mineworkers, and on the first Monday in May the Coal Board introduced it. The incentives to the miners were enormous. If they worked a week's full five shifts – which weren't increased – they received the sixth day's pay as a bonus (pieceworkers an additional 16 per cent of their earnings). In addition, overtime rates were improved (time and a half during normal shifts, double for Saturday afternoon to Sunday night) – and paid statutory holidays were to come.

The new beginning of January had been badly hit by the weather. The five-day week was a chance to begin again. All sides were full of optimism. 'This agreement places the British miner in the best position of any in the world,' said miners' leader Will Lawther. For his part, Manny Shinwell expressed hope that the new deal would reduce wilful absenteeism and bring unofficial strikes to an end.

Wilful absenteeism did reduce, by about half: men thought twice about taking a day off – doing so effectively cost two days' wages; nonetheless it hovered around 7 per cent for all workers and 9 per cent for face workers. But there was no reduction in unofficial strikes, which continued to erupt sporadically,[3] and an exasperated Shinwell commented, for the most trivial reasons; in many pits, he added, they were 'very frequent' and in some 'almost chronic'. This was especially the case in Scotland where he closed two pits because of constant disruptions.

August proved an ugly month. The first miners since nationalisation were dismissed for persistent absenteeism. Some collieries objected to working with the Poles, the majority ex-servicemen who didn't want to return home, saying their 'status was being degraded'.[4] Ten thousand pit clerks struck, complaining that the NUM wouldn't allow their union to take part in joint negotiations with the NCB; 6,000 overmen, deputies and shot firers struck, complaining that overmen were now worse off than the other two categories, which were subordinate grades. And when the government suggested that miners should temporarily add half an hour to the working day, both to meet the unrelenting demand for coal and to ensure that the country didn't find itself in the predicament of the previous winter, the miners instead wanted Saturday working reintroduced on overtime rates in addition to the bonus payment – which, the NCB pointed out, was not only against the spirit of the five-day agreement but was being sought when the conditions of the agreement weren't being met. Restrictive practices hadn't been given up, in defiance of the leadership, which was making the standardisation of wages impossible. A minority of face workers had rejected a small increase to their stints, and more, traditionally used to going home early when they'd completed their quota but now asked to work on, refused, staying in the pit but sitting down until loose it. On top of all these issues, a lightning strike at Grimethorpe colliery in Yorkshire brought out half the collieries in the county in the biggest dispute in the industry since Porter at the beginning of 1944.

The stoppage, which lasted five weeks, began with 150 collieries walking out after their 21ft stints, by agreement between the NBC and the NUM, were increased by 2ft. By the end of August, other collieries were out in sympathy. The leadership urged the men back; 'Burn Will Lawther' appeared in big letters on a wall at the entrance to Grimethrope pit yard. When officials arrived at Grimethorpe to talk to the men, barely 600 of the 2,600 attended; on 2 September only 454 bothered to fill in ballot forms. By the time the Coal Board conceded that the Grimethorpe stints would stay as they were, forty-six pits were wholly or partially stopped, with 55,000 men idle. *The Times* ruminated on the 'unreasoning loyalty' of miners.

Hundreds of Bevin Boys were caught up in the strike. Many went fruit picking on local farms to tide them over – or they went home. At Yorkshire Main, Les Thomas and most of the hundred other Bevin Boys there tried to go to work, but were turned back. 'In fact I tried three times,' Thomas says. 'I walked across the pit yard but each time two large gentlemen persuaded me it wasn't in my best interests.' Like most of the 500 inmates of the Sandringham Road hostel in Doncaster, he left until the strike was over, 'going home to a pal's in Cheshire. We biked over the Pennines, 70 or 80 miles I should think. Took us from 11 in the morning to midnight. We weren't half hungry!'

No sooner had the Yorkshire miners returned than 10,000 Scottish miners were out because of a strike by on-cost workers (underground specialists other than colliers) claiming a wage increase in line with firemen and overmen. There seemed to be no end to the unrest. Commentators were baffled by what the miners really wanted. Nationalisation ensured them of greater security and prosperity. Mechanisation was going into pits where there'd been none and was being upgraded in others. The old rope haulage methods were being replaced by electrically run conveyor belts, magnetic telephone systems were being installed at pit junctions, permanent electric lighting was becoming the norm. There was money for pit props, and existing conveyor belts were no longer having to be patched up. And a programme of installing first-class first aid facilities ('Our medical centre,' says Dennis Fisher, 'was just like a miniature hospital.'), pithead baths and canteens in every colliery without them was in hand (though it would take five years to complete). To many people it appeared that mining was unable to shake off the past, and the spiral of unrest and confrontation might be fateful.

Like many Bevin Boys, Brian Folkes watched the miners, and thought disillusionment was largely to blame:

> Most had naïvely believed that nationalisation was going to give them worker power, that they would be running their industry. It was sad to see them gradually realise that was a pipedream. The same old managers were there in charge of the pits, just as they'd always been. And there was now one union, which was making decisions remote from local issues – more remote than the old regions unions had been. The control was little different from what existed during the war.

A Cabinet reshuffle in October sent Manny Shinwell to Labour and National Service, Hugh Gaitskell taking over Fuel. At the end of the year he announced figures that clearly showed the majority of miners weren't part of the unrest, however widespread it was, and were trying hard to make nationalisation work. Overall, output a man-shift had reached 1.14 tons in the first week of September, the highest since April 1940, and, on the coalface, 2.96 tons, passing the 1939 level. The upshot was that the country entered the winter with over 6 million tons more coal stockpiled than in 1946. But that, Gaitskell warned, was no justification for 'unrestrained optimism': the five-day week, which was supposed to deliver not just the 18 million tons a year lost by the ending of the Saturday morning shift but an extra million a month into the bargain, had delivered only half; and a proportion of what was now in the coal yards was due to 600,000 tons of imports (equivalent, as it happened, to what was lost in the Yorkshire strikes).

The government was still trying to get more men into mining, without much success (an Admiralty message to the Fleet offering conditional release to miners serving as naval ratings or Royal Marines met virtually no response), but the newspapers were becoming exasperated with the miners, who continued to blame all their problems on a labour shortage. When the miners were awarded a 15s increase on the minimum wage, the newspapers accused the government of being conciliatory. When the government issued three-quarters of a million army food packs to the miners it was criticised for thinking they were more deserving than the old or families with young children.[5]

In view of everything that had happened in 1947, Bevin Boys may have been dismayed but hardly surprised by the announcement that as a safeguard against another bad winter, their releases were being suspended until the spring. No mass protest of the kind seen two years earlier ensued. The Bevin Boys just got on with it.

* * *

Every morning Joe Geohegan cycles from his home in New Brighton to Seacombe where he stores his bike and catches the ferry to Liverpool. Tram to Lime Street railway station. The 6a.m. train to Lea Green near St Helens. The walk to Sutton Manor colliery. Down in the cage for the 7a.m. shift. This is Geohegan's daily routine until he is released, one of the very last Bevin Boys compulsorily made a miner by the final, implemented ballot in April 1945 – but not one of the last Bevin Boys out. He leaves in March 1948; others will be in the pits until October.

Three years, give or take, was what most Bevin Boys averaged, though some found themselves in the pits for four and a half. Again, Bevin Boys were in fate's hands as to when the army group to which they were assigned came up, as indeed were the soldiers in those groups: age and length of service were the main criteria for military release, but it also depended on where units were and what they were engaged in.

The year 1947 was a curiously two-paced one for Bevin Boys. The excitement proffered by nationalisation and the snow made time pass quickly; from the spring it dragged, other than for some during periodic industrial unrest. Now it was a matter of sticking to the routine, and waiting for their number to fall due. It was almost worse when it did: the lag between it being posted and then arriving was indeterminate and seemingly interminable. Harold Gibson got notified on 4 April 1946 that his release group number was sixty-three, but it didn't come up for nearly eighteen months. 'I made a right nuisance of myself at Accrington labour exchange,' he says.

> I was down there all the time. 'Is this the week?' They got to know me. I'd hardly be in the door and someone would shout out, 'No, it isn't this week.' What was interesting about my little visits was that I ran into other Bevin Boys from Huncoat [colliery] asking the same question. Until then I'd thought I was the only Bevin Boy in the pit.
>
> It was the 12th of December 1947 that I was told the day I was going – the 19th. What a Christmas present that was.

There were five more groups behind him, waiting.

Some men were released ahead of their groups for various reasons. The scheme announced by Bevin in 1945 that let teachers (in addition to miners and building workers) go early from the forces was extended to Bevin Boys and that made civilians of a very few, including Eric Gulliver from Clifton colliery, in Nottingham. Jack Garland became a civilian eight months ahead of time, not because of the hernia he'd developed tumbling tubs down Markham pit near Bristol but because of the wait for the operation to repair it, during which he returned to his old office job on the understanding he'd be back, but his demob number came first. Whether or not a Bevin Boy got away ahead of his time was hardly of general interest, but it became so in the case of Gerald Smithson, created a national furore and was even debated in the Commons. The reason was that Smithson, who'd been at Askern colliery in his native Yorkshire since November 1944, had suddenly become famous – chosen to play cricket for the MCC against the West Indies.[6]

Even though he was due for release in January 1948, the month after the team were to sail and less than two months before his time was up, there were furious letters to *The Times* from those outraged that he might be permitted the concession of an early release. 'Coal is a more valuable export than cricket,' wrote one. Others took the opposite view, pleading with the Minister of Labour to make 'a human and generous' gesture. George Isaacs did let Smithson go but not, apparently, generously, granting only 'leave of absence on condition he completes his time on his return'. It was never realistic to suppose Smithson would do that and he didn't.

At least two Bevin Boys who were officially released and then told to go back didn't do so either. Both, one ex-army the other ex-RAF, received letters from Isaacs' Ministry within days of their leaving, informing them that their transfer out of the forces into the pits had made them liable to another six months beyond their group release. Once the newspapers carried the story the decision was overturned.

Life in the hostels changed as men left. Gradually the smaller hostels became uneconomic and closed, their inmates moving to others. There was an influx of newcomers in some of these: a few conscripts who'd chosen the pits over the forces,[7] a lot of Irish, even more East Europeans – not just Poles but Czechs, Latvians, Estonians, Lithuanians, some of them professional men and, as Dennis Fisher remembers at Chilton colliery in County Durham, 'all of them good workers, hard workers, and friendly. But we couldn't pronounce their names so they were all called Joe'.

A lot of the displaced spoke little or no English, but the Poles who took up residence at Oakdale hostel in south Wales, according to Geoff Rosling, 'had been training in Scotland and what English they spoke was in a Polish-Scottish accent, which many of the local Welsh girls found a fascinating combination, and those Bevin Boys who had earlier enjoyed their company now faced considerable competition'.

Many Bevin Boys found the hostels, so long their homes, less pleasant places than once they'd been. In some where they become a minority, the menu changed to suit East European tastes and proved too greasy for British palates. Worse, the newcomers had less respect for property, pit dirt was on the furniture, the washbasins were often filthy and the lavatories clogged. There were tensions in the summer of 1947 when the terrible winter, which had made fridges of the Nissen huts, gave way to a glorious summer, which made ovens of them: occupants sweltered and tempers were short. A few fights broke out, the majority among the Irish.

When they weren't fighting, most Bevin Boys found the Irish a source of endless entertainment, including Norman Brickell in the Dodworth hostel in Yorkshire who kept company with the three that moved into his hut.

> These three were older than us, from Enniscorthy in County Wexford, and the stories they told about some of the characters back home and the pranks they got up to kept the rest of us in stitches. We used to play cards, often later than we should have. On Saturday night I used to visit the Alhambra in Barnsley with them to enjoy the 'Liffey Water' [Guinness] they drank.
>
> Coming back was always an experience. The closer we got to the hostel the more contrite they became. And whatever state they were in they'd be on their knees praying. The thing we couldn't understand was why they did that only Saturday nights. And come Sunday nights their sinning would start all over again.

However much Brickell enjoyed the company, he did finally become disenchanted with Yorkshire. Soon after Hugh Gaitskill paid a visit to Monckton colliery ('He spent all of 45 minutes underground, poor man, though no doubt a few whiskies in the boardroom expunged the experience') Brickell packed up and took off without telling anybody. He didn't desert – he just wanted to be nearer home in Dorset,

> so I borrowed my dad's car and drove down to Somerset, the nearest place with mines. I went to the office at the Ludlow pit at Radstock and said 'I'm a Bevin Boy and I want a job', which I got. It was the worst day's work I've ever done. In Yorkshire the miners used to tell terrible stories of conditions in the bad old days. Well, Ludlow was like that, no mechanisation, no pit baths, nothing. I stuck it for six weeks then moved to Norton Hill at Midsomer Norton – all manual, but a better pit.

Moving round as he did, the authorities apparently lost track of Brickell. Conscientiously he soldiered on until one day in the spring of 1948 he suddenly decided he'd had enough, 'went back to my lodgings, got my gear and left. I didn't even bother collecting my wages.' Very likely he'd missed his release date. He never heard another word.

As the Coal Board rationalised the way the pits operated, it set higher targets for each colliery. Pleasey's in Derbyshire was 12,000 tons a week and when it was achieved the NCB's blue flag was hoisted. 'It didn't happen every week but certainly at intervals,' says Geoffrey Mockford. 'For the previous two years or so my loader end had filled 400–600 tubs a shift; this was now 800–1,000. We worked twice as hard, had longer hours and little snap time.' Most Bevin Boys, having won the right to think of themselves as experienced miners, took it in their stride. 'To be honest,' says Les Thomas, 'those of us left at Yorkshire Main, and that was most of us, had a kind of perverse pride in our professionalism.'

> You had to pay for your keep, but a lot of us did overtime. I did afternoon shift 2 to 10, back Friday night 11.30 to 3, in Saturday morning – the reintroduced Saturday was time and a half. To be honest, it was all right. The social side was still good – we were one of the hostels that was topped up from others. There was something wrong with you if you couldn't have a good time surrounded by 500 other lads.

Bevin Boys still took a day off here and there: 'a general', as Peter Archer remembers such a day being termed at the Old Coppice in Staffordshire; but Bevin Boy absenteeism was nothing like as common as in 1946. Brian Evans did absent himself for a month in the summer of 1947 – but that was with the NCA's blessing.

> The first World Youth Festival[8] was being held in Czechoslovakia and the Coal Board was sending eight representatives, one from each of the eight divisions. I knew the Midland area secretary and I convinced him that a junior reporter on the *Stourbridge Press* was exactly the person who should go – I'd be able to report back in detail. So I was the only Bevin Boy on the jolly.

However prevalent absenteeism had been or was, a proportion of Bevin Boys never deliberately absented themselves. Peter Archer for one; Harold Gibson at Huncoat near Blackburn another, though he would have if it hadn't been for his mother.

> I was living at home and my mother, who was strict nonconformist, ensured I never took a crafty day off when I was tempted. In a blizzard in '47 she made me wear a flying suit thing my brother brought back from the RAF. So I ploughed through three feet of snow and, of course, there were no trains. And she made me go in the Monday morning of my last week when I had flu and was in a real state. The bath attendant said 'You don't look fit, lad' and sent me to the under-manager who sent me home. Mum couldn't argue with that!

Indeed some Bevin Boys kept a completely clean sheet throughout their service. The newspapers dubbed Fred Cullen from Tottenham in London 'the perfect Bevin Boy' when he left Fernhill colliery in the Rhondda; the twenty-five-year-old tool-setter, who'd worked on the coalface, had never lost a shift in over three years. But there were other, unheralded, 'perfect Bevin Boys'. Among them was the

Jewish refugee Meir Weiss, who volunteered for the pits out of altruism and never missed the cage at Ashton Moss in Lancashire because he believed he was helping to win the war and then helping the country back on its feet after it. And Owen Jones, who modestly says his record at Welbeck in Nottinghamshire was so good 'because I was simply doing as I was told – and I was never sick or injured'. And Alan Brailsford at the Handworth in Yorkshire, whose record extended to nearly four years. 'A matter of having good health,' he says.

> Actually, I had a poor start in life. When I was born my grandmother said I wouldn't survive and I had everything going including double pneumonia as a child; I didn't start school until I was 7½. Thereafter I was remarkably fit and at 88 remain so. Yes, a matter of having good health. Though I would add without, I hope, being sanctimonious, I was also principled.

Most Bevin Boys made a big effort in the last week of their service as a matter of personal pride, including Des Knipe (who'd moved from the Derwent pit in Durham, closed by the NCB as uneconomic, to nearby Crookhall). His attendance record was hardly perfect, but being in at the end mattered to him, 'just for the wonderful feeling of being able to tell myself: "This is my last day."'

Denys Owen remembers the end of his very last shift vividly: he met the manager of Gresford colliery for the very first time – to be threatened with a charge of theft.

> As the shift ended, the kindly stationary engine driver asked if I could use some old engine grease. As it happened I could: my parents had moved into a large house where the outside door to the front porch was strangely connected by a chain to the inner, so that both doors opened simultaneously. And the chain was rusty. So I said yes and the chap filled my snap tin with the old grease. And, would you believe, I was searched and sent to the manager. A pretty unfortunate first meeting. But he relented and after giving me a severe dressing down let me go into the freedom of a dark and rainy night.

The two friends with whom Owen had gone right through service, who'd come off shift with him, were waiting uneasily. It was with a sense of occasion that Owen got someone to take their photograph, together in the pit yard, on his father's Kodak Brownie. Sadly, 'it came out black all over. Just like us.'

At Norton in Stoke, Geoff Baker, determined to go home with a full wage packet, turned in for a full final week, despite the miners he worked with doing their best to dissuade him. 'They'st worked dine 'ere fer nearly foer years,' they said, 'an' apart from a few bosted fingers and scars on thee back ter give thee a few colliers' marks, they'st come through this lot unscathed. So call it a dee anna dunn tak' any risk.' Baker refused to listen – and, superstitious to a man, the gang refused to have him on the coalface, Big Joe included. Baker worked out his week on haulage.

Superstition kept Geoffrey Mockford out of the Pleasley pit for his entire last week.

I kept thinking: 'Only one week to go.' I'd nearly been under a huge rock fall near the end of my time and I couldn't stop thinking about it. The idea formed in my brain that it was stupid to risk my life for one more week. I'd been lucky before; I mightn't be a lucky next time. I didn't go in the last five days.

I did go in on the last Friday to collect my final pay packet and some of my belongings in my locker. And I just walked away from Pleasley with a great sense of relieve and happiness but an emptiness, too, a regret that I didn't say goodbye to the men I worked with for more than three years.

Other Bevin Boys carried away happier last memories. Harold Gibson, down with flu during his last week, dragged himself into Huncoat on his last day to collect the week in hand the colliery owed him, but additionally was paid for the week he hadn't worked. 'That was unheard of,' he says. 'To me it meant they thought I'd done a good job, and that meant a lot. I certainly thought I had.' It meant a lot to Ronald Griffin when he left Baggeridge near Wolverhampton to be told by the manager: '"The men underground thought the world of you. You were a good worker." I was so touched by that I almost cried. I didn't think they thought anything of me at all. That was one of the most marvellous things that was ever said to me.'

The miners at Handworth near Sheffield gave Alan Brailsford a party before he departed – 'well,' he amends, 'a drinking session'. It was, he adds,

with some trepidation that I went along, knowing how these fellows could drink: eight pints at Saturday lunchtime and 'We'll go out and really have some tonight.' However, the session at the Greyhound at Attercliffe and another pub, the name of which I can't recall, went off successfully, although I must admit I couldn't manage to read the destination boards on the tramcars afterwards, so I walked all the way home to Broomhill.

The miners' ingrained superstition was something that Geoff Baker had come to understand well, but he was hurt, nevertheless, by his fellow face workers refusing to have him with them in his final week. It was with a sense of deflation that on his last day he handed in his lamp, showered and changed, and went to the canteen for his usual pint of tea – and 'became aware that there were a lot more men than usual'. Shyly, with a few halting speeches, Big Joe and company gave him a smoking cabinet for tobacco and pipes. 'I treasure that cabinet to this day and give it pride of place in the drawing room of my home,' Baker says. 'There's a small brass plate on it inscribed: *To Geoff from his mates from Norton Colliery*. It says it all – mates. If they'd put colleagues it wouldn't have seemed appropriate.'

Miners had their ups and down with the Bevin Boys over the years; in the final analysis, however, the majority of those who served out their time had the miners' respect, even in south Wales where for many reasons Bevin Boys integrated less well than in other parts of the country. The following letter – typical of others written elsewhere – was sent by the South Wales Area Council of the NUM Tirpentwys Lodge to Bevin Boys as their release came due, and is clearly heartfelt:

Dear Comrade

On behalf of the Officers and members of the above Lodge, it is my pleasure to convey to you their sincere thanks for your assistance in a difficult but interesting period of our history. You came in at a time when we were bumping along the bottom and when we took the upward turn, owing to pressure of other work, we haven't been able to pay the attention to you that most of us desire, but that did not prevent us from keeping an eye on the manly way in which you faced up to the problems in strange surroundings.

We sincerely hope that you will look upon the time spent with us as one phase in life that steeled you with the confidence and courage to face up to anything that may arise in the future.

If after your present break you decide to return to mining as a life's calling, you will be heartily welcome, if you decide otherwise, we shall retain memories of the 'Bevin Boys' mainly of a pleasant character.

Wishing you all that is best in the years to come, and that we shall again have the pleasure of meeting you in more favourable surroundings.

In his address to the House in 1945, William Teeling said he thought that no more than 1 per cent of Bevin Boys were likely to stay in coalmining; later, union leader Will Lawther thought perhaps 5 per cent might. A number of men in this narrative were among those who did or left and then returned (see Biographies pp. 218–24).

As their release fell due during 1947, the Coal Board also wrote to the Bevin Boys and in fulsome terms, but, it has to be said, with an ulterior motive:

We thank you for your service to the Coal Mining industry and appreciate very much the good work you have done. You have become somewhat experienced by now and we shall be sorry to lose your services. You came to us at a time of great stress and strain; the need for coal and for manpower is as great, if not greater, than when you came to us. We are therefore appealing to you to consider favourably remaining with us for a further temporary period to help tide us over the immediate crisis. We can definitely promise you this it will not prejudice your chance and opportunity for release if you decide to remain and help us out in this Emergency. What we are really asking is for you to defer your application for release for a further temporary period.

Peter Archer responded affirmatively and stayed for several months beyond his time. 'To be honest it was less altruism than because I had my own agenda,' he admits. 'I'd already completed my BA. I timed my leaving to start my law degree at university.' He was one of the few Bevin Boys who even considered saying yes. 'My reply,' says Brian Folkes, 'wouldn't bear repeating in polite company.'

Bevin Boys who, one way or another, had escaped the pits and gone into the army or air force (none made it into the navy) gradually got their release, too, like everyone else. Warwick Taylor went home to Harrow from RAF Wharton in Lancashire, his discharge papers highly commending him for his qualities and his conduct but, at the same time, sending a shiver down his spine in saying: 'Can be recommended to return to his own job in coalmining.'

The letter that Bevin Boys received from the Ministry of Labour giving their release group number and date of leaving was eagerly awaited, but it gave information that was less well received:

You may, if you wish, take up employment in any industry during the period of 56 days immediately following the date on which you leave coalmining employment. Thereafter you will be subject to the restrictions in seeking and obtaining employment imposed by the Control of Engagement Order in the same way as other civilians, and you should, if you are then unemployed, attend with this letter at a Local Office of the Ministry for advice regarding your position in relation to the choice of further employment . . .

If after leaving coal mining you claim unemployment benefit, the fact that you are released from your obligation to serve in the coal mining industry may not, of itself, enable you to satisfy the Statutory Authorities that you had just cause under Section 27 of the Unemployment Insurance Act 1936, for voluntarily leaving your employment. You therefore are advised to make sure that you have a definite prospect of alternative employment before leaving coal mining.

The second paragraph was clear enough: as far as the government was concerned, in leaving the pits Bevin Boys weren't so much being released from service as making themselves deliberately unemployed, and if they failed to get a job, they couldn't expect any help from the State. Having in most cases given three years or more of their lives to their country, Bevin Boys considered that the height of ingratitude. But the first paragraph was more ominous: it seemed to contain the veiled threat that if they didn't get a job within two months they could be sent back to coalmining. That can't have been the Ministry's intention and many saw it as the piece of clumsy officialese it undoubtedly was. Observes Geoff Darby: 'After all the hard work, colleagues killed or badly injured, sustaining my own injury – and then we got that bureaucratic rubbish. And so say all of us!' But Bert McBain-Lee was one of those who thought the Ministry's threat was real. 'I wasn't taking any chances. I took the first job that came my way – salesman for a general goods warehouse. I was scared of being directed back to the pit.'

Bevin Boys had other grievances. The government had tied them to the army for release, but treated them very differently. Ex-soldiers received a gratuity (dependent on their length of service) and their first fifty-six days out of uniform were paid leave; and if at the end of it they weren't fixed up with work, they were entitled to the unemployment benefit the Bevin Boys were denied. Ex-soldiers also came out with a civilian suit, a raincoat, a trilby or cap, a shirt, two pairs of socks and a pair of shoes and, in addition, 120 clothing coupons. None of these things came the Bevin Boys' way and Geoffrey Mockford was angry about it. 'It was so petty,' he says. 'How much would it have cost the government to kit us out to start a normal life? The difference between the soldiers and us was that they hadn't worn their own clothes to do their job. I'd used up everything I owned down the pit.'

A significant entitlement that all ex-servicemen and women enjoyed was reinstatement in their old job, if it existed (many, of course, didn't). But Bevin Boys had no rights in the matter, as Harold Gibson discovered. Notified in April 1946 of his release day the following year, he immediately applied for his job back in Blackburn tax office. It was seven months before he got a reply from the Inland Revenue Accountant and Comptroller General, which told him: 'I have to refer to your letter and with regret have to advise you that persons directed for service in Mining are unable to claim reinstatement within the terms of the Reinstatement in Civil Employment Act 1944.'

All, however, wasn't lost. Gibson had taken the precaution of sitting the Civil Service Reconstruction Examination the previous month, having attended night school and then often going straight from it on to the night shift. 'It was a hard slog,' he says. 'My maths was fine, but trying to write essays after two years underground was bloody hard slog.' He passed – 'without the 50 bonus marks given to forces personnel. If I'd failed and come within 50 marks of the pass I'd have been outraged.'

When it came to examinations, those coming out of the forces were looked on favourably by the professions. Brian Folkes found exactly the same thing in surveying.

I'd put in a big effort to pass the first RICS exam while still a Bevin Boy, whereas my contemporaries returning from the services were granted full exemption from that. And they were allowed a 5 per cent lowering of the pass mark for the subsequent two. As a Bevin Boy I got no concessions at all.

Getting a job wasn't easy in post-war Britain: 5 million men and women were coming back from the forces; 4 million were leaving various forms of war work. Folkes didn't get his old job back ('they'd been doing a lot of stuff with airfields and hospitals in the war, work which naturally had dried up, so they didn't want me back'), but eventually went into house surveying. One day he ran into the president of the RICS when they both happened to be conducting a survey on the same house and Folkes complained to him about the situation. 'In some ways I was lucky to pass the first exam,' he says.

It wasn't easy coming up tired from the pit in your pit dirt and getting down to the books. A lot of servicemen at the latter end of the war had an easy time of it. Their case was certainly no better than a Bevin Boy's. And as I told the president, 5 per cent can be critical in an exam. Pointless, really, but it irked me.

The universities were no less charitable to returning servicemen, as Alan Gregory found when he went up to Cambridge to read classics and there were 'blokes there from Dulwich who'd never been academic and had scraped through the Higher School Certificate. They hadn't got scholarships never mind exhibitions, but there they were – allowed in just before they'd been in the forces. The beginning of the lowering of educational standards, in my opinion.'

The one entitlement the government felt able to allow Bevin Boys was the right to resume interrupted apprenticeships and to obtain educational funding – 'a right,' Ness Edwards, the Ministry of Labour's Permanent Secretary, said in the 1945 Commons debate on Bevin Boys' demobilisation, 'not conferred on the ordinary worker'. A number of Bevin Boys availed themselves of the opportunity including Phil Robinson, who got a diploma in diagnostic geology at Wigan mining college, and Ian McInnes, whose 'serviceman's grant saw me through University College, Nottingham, and gained me a mining degree, and a first-class manager's certificate – and my three years as a Bevin Boy counted against that, otherwise it would have taken five years'.

In the parliamentary debate, the question of a medal for Bevin Boys had arisen and was quashed by Edwards: there was to be no medal for Bevin Boys or, for that matter, miners: 'Why should a miner have medals more than a gun maker?' Edwards wanted to know. 'If we are to give medals, we must give them to the whole civilian population, and not least to the housewives of Britain. I do not think that is a tenable proposition, and we cannot treat the Bevin Boys differently from the rest of the community.'

The government's position was undoubtedly necessary and fair: in logic and in law the Bevin Boys were civilians. But almost to a man, whether they'd left the armed services (at the government's behest) for the pits, or gone down them as ballotees or as conscripts who'd volunteered themselves, the Bevin Boys disagreed. In 1944, when his appeal against the ballot had been turned down, Harold Gibson had written to his MP who replied, saying that 'no sailor, soldier or airman has a more important job than the miner'. 'We were the equal of the sailor, soldier and airmen when it suited them,' says Gibson. 'It was a different story when they'd done with us. The irony of what the MP wrote wasn't, shall we say, lost on me.'

Dennis Fisher, son of a miner who'd became a reluctant miner but remained a miner when his Bevin Boy days were over, feels exactly the same.

I reckon I'd done my bit and done old Bevin well. All we wanted was a bit of alloy and a ribbon to wear on our chests. The way we were treated left a bitter taste. If I had to do my bit all over again and at the end of it receive no recognition for services rendered, then the government would have to import peat from Ireland to keep the home fires burning.

In his impassioned address to the House in 1945, William Teeling called the Bevin Boy scheme 'an experiment [that] has on the whole turned out to be a failure'. Many Bevin Boys agree and Brian Folkes speaks for those who feel that in saying, 'It was socialist dogma triumphing over practical commonsense – there was no way that such an ill-assorted bunch, mostly from totally alien backgrounds, could make any real impact on coal production.'

But that seems a harsh judgement, which, perhaps, is coloured by what happened after the war, when the Bevin Boys received the psychological blow of being told they were being kept in the pits, after which a sizeable proportion refused to pull their weight or simply disappeared. While the war was on, George Isaacs as Minister of Labour was able to tell the House that 'the Bevin Boys are doing an invaluable job . . . we cannot do without them'. 'Some say we helped win the war, a lot say no,' comments Harold Gibson. 'All I can say is that I think I did a good job and there are plenty of Bevin Boys who think they did too.'

As far as Peter Archer is concerned, the failure wasn't in the Bevin Boy scheme but in 'the government's lack of imagination' in not realising that the miners were more important in the pits than on the front line. 'The government created a hole in the dyke,' he says. 'It was left to the Bevin Boys to plug it.' And, whatever qualifications have to be put on their contribution, the Bevin Boys did plug it. By taking over the haulage jobs, they freed over 11,000 miners to work on the coalface – where, indeed, between 6,000 and 7,000 Bevin Boys were of the stuff to work there themselves. What is indisputable is that the Bevin Boys stopped coal output falling further and faster than it did. There may have been times when it ran close, but Britain's war effort never stalled because of a lack of coal.

It's improbable that any Bevin Boy during his time in the pits met the architect of his fate. Norman Brickell almost did, when in April 1945 he read in the *Yorkshire Post* that Bevin was visiting Leeds on the following Saturday. 'I admired him as a forceful politician, and I thought I'd like to talk to him,' he says. 'After all, he was the guy responsible for me being in the pit.' Brickell caught the train from Royston and sat on the town hall steps – his one and only visit to Leeds.

City dignitaries and police began to arrive and prompt at 2pm a line of official cars drew up and Bevin stepped out. As the phalanx of policemen around him climbed the step I thought 'This is my chance' – I was one of his Boys, wasn't I? And I was pushed aside. Me and my naïvety. And so much for the special relationship I thought I had with the man who drew my number out of his hat.

When their numbers came out of Bevin hat, few Bevin Boys felt as amenably disposed towards the Minister of Labour as Brickell was, at least before he got to Leeds. Many would have agreed with John Platts-Mills – who knew Bevin personally – that he was 'a horrible fellow . . . a chap to be despised on principle' (though Platts-Mills admitted Bevin's 'idea of forcing people into the pits was a sensible one because we were dependent on coal').[9] Reg Taylor remembers going to the cinema with a group of Bevin Boys during training in Pontefract and when Bevin appeared in the newsreel 'Uproar broke out. Howls of derision, catcalls, rude words, whistling. The cinema manager stopped the show until we quietened down.'

Time has mellowed most men's opinion. Albert Mitchell (from Bristol, where Bevin was once a drayman) merely says: 'He did what he had to do.' And Phil Yates, when asked what he thought of Bevin, by way of an answer breaks into 'the Bevin Boy song', knocked up sixty-odd years ago around a piano in a pub or a hostel (though no one knows where or by whom), sung to the tune of 'Way Down in Dixie' from the Laurel and Hardy film *Way Out West*:

We had to join, we had to join, we had to join old Bevin's army,
Three quid a week and bugger all to eat,
Hob-nailed boots and blisters on your feet.
We had to join, we had to join, we had to join old Bevin's army,
If it wasn't for the war we'd be where we were before,
Bevin, you're bloody balmy!

But a man who remains as angry at Bevin now as he was in 1944 is Dan White. 'Every war has a complete idiot,' he says vehemently. 'It was Haig in the First War – well, Bevin came out of the same mould. I would have enjoyed giving him a good kicking. I wouldn't have pissed on him if he was on fire in the gutter.'

Many Bevin Boys remember a poster that hung in their pit canteens, which under a photo of Winston Churchill displayed a quote from a speech he made in Westminster Central Hall in October 1942:

> We shall not fail, and then some day when children ask: 'What did you do to win this inheritance for us and to make our name so respected among men?' one will say: 'I was a fighter pilot'; another will say: 'I was in the submarine service'; another: 'I marched with the Eighth Army'. A fourth will say: 'None of you could have lived without the convoys and the Merchant seamen'; and you, in your turn, will say with equal price and with equal right: 'We cut the coal.'

It would be to quibble to say that Churchill was addressing the miners before the Bevin Boys went down the pits. To Churchill the miners, whatever their faults, were a symbol of Britain's indomitable spirit. The Bevin Boys, ex-servicemen and ballotees, optants and volunteers, whatever their faults, were no less miners. Churchill, surely, would have wanted his words to embrace them too.

Epilogue

Boys Forever

By the end of October 1948 the Bevin Boys were all gone from the pits, save those who'd elected to stay, or return, including some of the men in this narrative and others, like Jock Purdon, the poet-folksinging miner who, when he wasn't performing, spent his life down a pit in Chester-le-Street, County Durham. The rest returned to their lives: ordinary lives for the most part, famous lives in a few cases, like those of Eric Morecambe and others who've appeared in these pages; or eminent lives, like those of the physicist Sir Martin Wood, and Members of Parliament Geoffrey Finsberg (Conservative) and Peter Archer (Labour and, like the remarkable John Platts-Mills, a QC), both later elevated to the Lords.

With time, the majority of balloted Bevin Boys came to believe that they were lucky to have been denied the forces. As Phil Yates observes, 'Mining was a risky business, but I wasn't shot at or shot down. Would I have survived as a ship's stoker or if I'd taken part in the D-Day landing?' Many Bevin Boys, however, found they preferred not to talk about the war, particularly to men of their age, because it inevitably raised questions about what they'd done during it. Some felt a sense of inferiority for not having been in uniform. 'For years after the war the topic would came up all the time,' says Dan Duhig, 'and you'd listen to blokes: I was in Egypt, I was in Italy, I did this, I did that. I had the feeling I'd done nothing.' Dan White goes so far as to say he felt 'there was a stigma to having been a Bevin Boy'. What upset many, Syd Walker among them, was that if they did say they'd been Bevin Boys, the belief persisted that they were conscientious objectors. 'I met a man on a train who made that assumption,' Walker says. 'I put him straight and he apologised profusely, but I realised it would be many years, if ever, before people would know the truth about the Bevin Boy conscripts.'

For some Bevin Boys, an unpleasant legacy of coalmining was that it left them with bad dreams of being forced back to the pits, and worse, of being trapped in a closing. David Day suffered from bad dreams; so did Jack Garland; and Geoffrey Mockford. 'They were very frequent and very real to me,' Mockford says. 'Thankfully, now, such dreams are rare.'

In the decades after they left the pits, Bevin Boys watched the industry to which they once belonged with more interest than most and could only shake their heads as it imploded. The high level of local strikes with which they were so familiar continued, and escalated as pits were closed for economic reasons, leading to violent confrontations first with the Heath government in the seventies and then, terminally, with the Thatcher government in the eighties. At nationalisation, NCB chairman Lord Hyndley spoke of mining 20 billion tons of coal in the next 100 years. Within forty years, gas, oil and nuclear power had almost replaced coal and now only five pits, in private hands, operate, producing less than a third of the coal the country does use.

'I never thought I'd live to see the day when there weren't any pits left in my county,' says Durham's Dennis Fisher. 'So now we import coal from places like Columbia, Poland, Austria and who knows where else. Ironic that at Newcastle there's a depot to store all this foreign coal, which makes the old saying about taking coals to Newcastle into a reality.'

But Bevin Boys were not and are not prepared to pass judgement on the miners. 'If anyone has criticised miners in my presence,' says Harold Gibson, 'I in no uncertain terms asked them if they had any idea of the miners' life and suggest that they exchange places for a while.' 'I would stand up for the miners against anybody who denigrated them,' says Owen Jones. Alan Gregory, who for twenty years until the late sixties worked for the Ministry of Fuel, 'had no patience with people who badmouthed the miners. I used to walk over from Millbank to Whitehall with a department solicitor who, one day, was criticising the miners to the point where I told him to shut up. He could criticise the idiocies of government and of the NUM, but I wouldn't hear a word against the miners – damn good chaps who were grossly misled by their leadership.' 'Led up a cul-de-sac,' is how Victor Simons sees it. 'I was working for the Coal Board in Yorkshire at the time of the year-long strike and once or twice had to face Arthur Scargill across a table. The arrogance with which he led the industry into chaos made me extremely angry.'

More than a few Bevin Boys stayed in touch with men they'd worked with, including Geoff Baker, who visited Big Joe until he died, in his late seventies, 'riddled with pneumoconiosis'; Big Joe never tired of introducing Baker to people as 'the bugger who wore an evening suit down the pit'. Baker named his house Joffers – Big Joe's nickname for him – 'as a kind of tribute to him: I loved that man.' On a visit to England from his home in Canada in 2006, Jack Garland went to see his old butty, then aged ninety-one.

Down the years many Bevin Boys have paid nostalgic visits to their old collieries. In 1975 Owen Jones wrote to Welbeck, asking if he could again take the cage to the pit bottom. 'I was presented with a helmet and a miniature miner's lamp, and walked into the pit yard to shake hands with many of the men who had known me.' Twenty years later, as coalmining became a thing of the past, there was little or nothing of the collieries to see. Doug Ayres found a motorway instead of the Homer and Sutherland; Harold Gibson found two clumps of trees where once the Huncoat shafts used to be – 'but the chip shop had photos galore on the walls'. John Cook's return to the Louisa was just as dispiriting, but his return to County Durham was nevertheless rewarding. He went into Durham cathedral and was 'thrilled to see that they had a miners' memorial book on display, and for coincidence it was opened at the page commemorating the Louisa pit disaster of 1947. And the names of both Bevin Boys were shown among the dead, though sadly there was no indication that they were Bevin Boys.'

For thirty-five years after they left the pits, the Bevin Boys dropped out of sight – though in May 1964, when Mods and Rockers were rampaging in Margate, William Telling suggested in the House that perhaps a dose of 'the Bevin Boys' might do them some good. Then in 1983, just ahead of the fortieth anniversary of the implementation of the conscription ballot, Radio 4 and Tyne Tees Television aired half-hour documentaries, prompting Stanley Durban, an old Bevin Boy from Croydon, to approach Chatterley Whitfield mining museum in Stoke, suggesting it host a Bevin Boy reunion. As a publicity promotion the museum in 1989 offered free admission to any Bevin Boy who cared to come along on a certain day – and were surprised when thirty showed up.

On the back of this, the Bevin Boys Association was founded and a small group began using the museum as a base to get publicity and recruit members. Warwick Taylor, who's already begun to research the history of the Bevin Boys to produce a permanent record, soon joined. In 1993 Geoffrey Finsberg, who'd served at Glapwell colliery in Derbyshire, was elected the association's first president. When he died suddenly three years later (Lady Finsberg taking over her husband's role) Taylor, by now the association's archivist, press officer, publicist and public relations man, also became vice-president.

When he began gathering material for his book, Taylor had his title in mind: *The Forgotten Conscript*. Largely due to his efforts, the title was becoming less apposite by the time he got into print in 1995. Information about the Bevin Boys was spreading – and now every mining museum has a Bevin Boy display; the association has a touring exhibition; and members all over the country are out and about giving lectures and going to talk to schools. The association has its own website and its archive is housed in the Imperial War Museum. Today, membership is nearly 3,000, extraordinarily having risen year on year, despite the inevitability of names having to be regularly added to the association's book of remembrance; realistically, with all its members in their eighties, membership must soon decline.

The recognition that the Bevin Boys were denied for so long has belatedly come their way. Their wartime contribution was first officially acknowledged in speeches made by the then Prime Minister John Major and the Queen in their addresses to mark the fiftieth anniversary of VE Day in 1995. Three years later they won the right to march in the Remembrance Sunday parade to the Cenotaph, previously blocked by the British Legion, thanks largely to the lobbying of Lord Mason, once a miner himself. And in April 2003, a delegation from the association went to the National Memorial Arboretum at Alrewas in Staffordshire – conceived in the late eighties as a living tribute to the wartime generations of the twentieth century – to plant three trees in tribute to all Bevin Boys: an oak for England, a mountain ash for Wales, and a spruce for Scotland.

Finally, in February 2007, after a long campaign, the government announced that the Bevin Boys were to get 'a bit of alloy' to wear on their chests – not the medal they wanted, true, but a badge, something tangible that proved, as Morry Pearce emailed everybody: 'We're officially WAR VETERANS now.'

Acknowledgements

Over the years the Bevin Boys Association through its newsletter has sought men's reminiscences for its archive. Most of the men in 'Biographies', which follows, made written contributions and I've been able to draw on these as well as interview their writers, all of whom I would like to thank. I have also drawn on the accounts of the following, now deceased, whom I can only thank posthumously: Peter Allen; Gerald Carey; John Cook; Roland Garratt; Ernest Noble; Stan Payne; George Poston; Jim Ribbans; Geoff Rosling; Len Ungate; Eric Ward; John Wiffen.

I owe a special debt of gratitude for the unstinting help of Warwick Taylor, keeper of the Bevin Boy flame.

Biographies

Since the first publication of this book in 2008 a number of the men who appear in it have died: Geoff Baker, Alan Brailsford, Desmond Edwards, Des Knipe, Albert Mitchell, Geoffrey Mockford, Roger Spencer.

Archer, Peter (Lord Archer) Having obtained his BA while still a Bevin Boy, became Bachelor of Law at University College and Master of Law at London School of Economics. Admitted to Gray's Inn in 1952 and practised until 1999. MP (Labour) for Rowley Regis; Tipton; and Warley West until 1992. A QC in 1971, he was a member of the Shadow Cabinet 1980–87 and created Baron Archer of Sandwell in the West Midlands in 1992. Lives Wraysbury, Berks.

Ayres, Doug London engineering student before call-up, joined the civil engineering laboratory of British Railways Western Region in 1949, working mainly on problems relating to geology and becoming head of section 1963. Appointed soil mechanics engineer of the amalgamated British Railway Board HQ section in 1966, from which post he retired in 1989. Still active as a geotechnical consultant. Lives London.

Baker, Geoff Pupil from St John's public school for the sons of clergy, on release from service ran a boys' club in Forest Gate, east London, then became deputy superintendent of a residential home for boys in Broadstairs, with his wife as deputy matron. Subsequently ran a mixed children's home in Herne Bay and then another in Sandwich, as well as having responsibility for eight other homes in Kent. Finally ran a home for the elderly in Deal. Spent twenty years as a magistrate.

Bates, Jim Revd Became a student minister in Huntington, Cambridgeshire, waiting for a place at Bristol Methodist Theological College. After ordination worked in Wells, Somerset, Ushaw Moor, Durham, and a boarding school in Canterbury, before spending twenty-seven years as the chaplain of Southlands Methodist teacher training college, Wimbledon, now part of Roehampton University. Lives in Ilfracombe, Devon. Lectures on world religion - when he isn't painting and drawing.

Bown, Ron Working in a factory making batteries and rectifiers for the navy before call-up, became a railway fireman, then a die caster, staying with the same company all his working life in Wolverhampton, still his home.

Brailsford, Alan Accountant in Sheffield steel manufacturers who after release worked for a wholesale business in Chesterfield, then as company secretary for a large American caterpillar dealership in Spalding before becoming London area manager in Windsor. Retired as company secretary of a large broiler chicken company in Hereford. MBE after a lifetime in scouting.

Brickell, Norman Printing apprentice in family business in Gillingham, Dorset. Returned to printing and worked for the same firm in Gillingham and then Shaftesbury until retirement. Lives Bournemouth.

Brinsmead, John Printer's apprentice from Worthing. Returned to printing, later owning own jobbing print business in his home town, where he still lives.

Brown, Tony Signwriter from Leicester who returned to his trade, specialising in advertising graphic artwork for trade fairs and exhibitions. Home, Welwyn Garden City, Hertfordshire.

Chislett, Stuart Office clerk from Epson, became export manager for a tyre company, then went to West Africa to work for the Cameroons development corporation in administration. Lives South Molton, Devon.

Da Costa, Ken Electrician from Cricklewood on call-up, for over thirty years ran own business as a manufacturer's agent selling upholstery and curtain material. Lives Stanmore, Essex.

Darby, Geoff Became a Bevin Boy from Huddersfield Corporation waterworks department. When he came out, was a youth club leader and worked for the YMCA before rejoining the corporation in the electricity department. After night school in electrical and mechanical engineering joined the Central Electricity Generating Board, working in Guildford, at Dungeness nuclear power station, and then in Leeds, where he lives.

Doorbar, Roy Butcher from Smallthorne, Staffs, he returned to his trade after leaving the local pit. Having managed several shops, he opened his own in Glastonbury, Somerset. He continues to live there.

Duhig, Dan From Poplar, east London, after service went to work for a printer in Oxford, packing books. Eventually got a job with the Pressed Steel Company,

Cowley, making car bodies, and worked there for forty-four years. Home, Botley, Oxfordshire.

Eddolls, Roy   Garage mechanic from Bletchingley, Surrey, returned to his trade and worked for the same firm in Oxted for thirty years before moving to the vehicle maintenance shop of a frozen food company depot in Redhill. Lives Oxted.

Edmonds, Michael   Trainee architect from Bere Regis, Dorset, went to work for the Coal Board in Cardiff after qualifying, designing pithead baths, medical centres and housing. Set up own architectural practice in London in 1957 and won a competition to design a 9m-wide ceramic mural at the pneumoconiosis research unit, Llandough hospital, near Cardiff, paid for by the NUM and the Medical Research Council. Lives Montgomery, Powys.

Edwards, Desmond   Welsh miner's son from Cwmcwmdare who spent a career teaching geography in Birmingham, Lancashire, East Anglia and East Sussex, the county in which he lived in Framfield, near Uckfield.

Etty, John   Semi-professional rugby player with his home town of Batley, Oldham and Wakefield; represented the British Empire against Wales. Spent forty-one years in local government, mostly in social services, in Yorkshire, Wiltshire and Lancashire. Retired as chief admin officer for Blackpool and The Fylde. Lives Fleetwood.

Evans, Brian   Junior reporter in Stourbridge before conscription, worked for the Coal Board PR department on release, then as a journalist on the Board's magazine, *Coal*. Later worked as press officer for a tyre company, an electrical furnace manufacturer, as advertising manager for a radio valve company, then as general manager of a horse racing newspaper. Finally ran own publicity agency in London. Lives Ickenham, Middlesex.

Faulkner, Dennis   Junior BBC radio engineer from Gloucester, became a radio fitter at a local RAF station after service, then went into cable television research. When cable collapsed (not to be revived until the 1990s), worked as service manager for an electronic company in Leeds and then as office manager of a property company. Lives Leeds.

Fisher, Dennis   Son of a Durham miner from Gurney Valley, Bishop Auckland, continued to work at Chilton colliery for twenty years until it closed in the 1960s. Worked part-time for an auctioneer's and than an antique dealer's until he got a job with an American engineering firm in the heat treatment department. Lives Newton Aycliffe, 5 miles from Darlington.

Fisher, Reg   Motor mechanic from Wembley. Stayed in light engineering and then instrument making until he was thirty-five. Taught technical illustration, became chief illustrator for a large lorry company, and then manager of a technical studio. For fifteen years before retirement ran his own commercial art studio in London. Still entertaining, staging singalong shows. Lives Iver Heath, Bucks.

Folkes, Brian   Trainee chartered surveyor from Buckhurst Hill, Essex, went into estate agency and valuation after the pits. Later worked for Pickfords, looking after the company property portfolio and acquiring new premises, then for Eastern Region of British Railways, finishing as estate surveyor. Lives Bulcote, 6 miles from Nottingham.

Garland, Jack   Clerk in a Bristol company making cardboard for a cigarette manufacturer, he qualified as an accountant and became a member of the Institute of Management Accountants. Worked variously for Reed paper group in Gloucester and Aylesford, for an aircraft group in Cheltenham and then for a subsidiary in Canada before returning to Cheltenham, working for a stationery firm as secretary and treasurer. Returned to work for Reed in Toronto and then Montreal. Lives Ontario.

Geohegan, Joe Revd   Caught in the last Bevin Boy ballot in April 1945, he left the pits expecting to return to his studies as a Roman Catholic seminarist, only to discover he had TB and spent fourteen months in a sanatorium. Returned to his studies – now in the Netherlands – and when ordained in 1955 went to Kenya, where he spent forty-one years in Nyanza Province. On his return, helped out in various Lancashire dioceses. Lives Formby, Merseyside.

Gibbs, Bill   Junior bank clerk from Grimsby, he was a policeman for two years, found the wage insufficient, and returned to mining at Clipstone colliery near Mansfield. After ten years bought a fish and chip shop in Cleethorpes, then started a mobile wet fish business in Grimsby. For the last twenty years before retirement worked as an operator at the Conoco oil refinery at Immingham. Lives Horncastle, Lincolnshire.

Gibson, Harold   Failed to get reinstatement to his job with Blackburn tax officer after finishing his time, but passed the necessary examination. Served in Chorley, Blackburn, Cardiff, Pontypool and Newport, retiring in 1986 as an inspector. Lives Croesyceiliog, Cwmbran, south Wales.

Gilbert, Arthur   After Bevin Boy service went home to Cheadle, Staffs, and joined the local copper products company, from which he retired as deputy head of sales. Still lives in Cheadle.

Gregory, Alan Dulwich college pupil from Forest Hill, who left Cambridge for the Civil Service and the Ministry of Fuel and Power, where he became a principal, then an assistant secretary in the petroleum, gas, finances and again the petroleum divisions. After twenty years he left to join BP. Lives Cobham, Surrey.

Griffin, Ronald On leaving the pit studied drama on a scholarship in Birmingham and was in rep for a few years before ten years as a youth drama instructor in Staffordshire and Worcestershire. Worked as a civilian accounts clerk for the Royal Army Pay Corps, then the RAF, then as a telephonist for the Post Office in Kidderminster and finally at Lea Castle hospital, Wolverley. Has written a number of novels and won a BBC World Service 'life story' competition with his Bevin Boy experiences.

Hitch, Bill Market gardener from Haddenham near Ely, went back to Cambridgeshire and worked on a poultry farm, but returned to the North East, working at the Nettlesworth and Sacriston collieries. Lives Sacriston, Durham.

Jefferies, Ernie Spectacle-frame maker from Elephant and Castle, returned to his previous employment, then worked in various factories become becoming a milkman for twenty years. Lives Kingsbury NW9.

Jones, Owen Apprentice cabinetmaker from Portslade near Brighton, who after release went to Plumpton agricultural college to study farm engineering. Worked on the one farm at Uckfield, East Sussex, for forty years. Still making models of steam engines, 'inspired by the wonderful steam engines at Welbeck colliery'. Lives Wells, Somerset.

Kelly, Cecil Gardener from Bolton. Emigrated to New Zealand in 1952 and there fifty-two years, always working as a gardener. Lives Thornton Cleveleys, Lancashire.

Kinnear, Bob Butcher's boy in Folkstone after evacuation from Catford, south east London, came back to work on maintenance of boot repair machines at the Royal Arsenal Co-op Society, Woolwich. After a career in the Metropolitan police, retiring as a sergeant, joined the Foreign Office security staff at embassies in Warsaw, Bonn, Cairo, Paris and Vienna. Later, was employed at the American Embassy in London. Lives Bromley, Kent.

Knipe, Des Electrical apprentice from Bexley, then a municipal borough outside London, worked on buildings sites all his working life. Lived in nearby Welling.

Lane, Alan Printing apprentice from Bristol, went back to complete his apprenticeship and remained in printing, in Bristol and Pontypridd, until he retired. Lives Portishead, Somerset.

Marshall, John   Radio technician from London who on release spent a lifetime travelling the world as a sound engineer. Lives in the village of Ellington on the Northumberland coast – where the pit closed in January 2005.

McBain-Lee, Bert   Insurance office clerk in Liverpool, became a member of the Chartered Institute of Marketing and spent most of his life in sales and marketing in the fast food trade, confectionary, and industrial materials. For twenty years while in Scotland ran his own made-to-order kilt shop in Dunblane. Retired as general manager of a holiday complex on the Isle of Man, where he lives in Castletown.

McInnes, Ian   Engineering student from Dover who obtained a mining degree from University College, Nottingham, and subsequently worked in Wales, Yorkshire, Argentina, Canada, South Africa, America and Australia. He retired as a mining consultant. Lives Peasedown St John, a former mining village 6 miles from Bath.

Mitchell, Albert   War factory worker from Bristol, went into the army Service Corps, transferred to the Catering Corps and served twelve years as a regular, leaving as a staff sergeant. Worked as a cook at a radar site in Northumberland, then returned to Bristol where he did numerous jobs, mostly driving, including for the Co-op.

Mockford, Geoffrey   Cabinetmaker from Babraham, Cambs. On release, worked briefly for the Ministry of Fuel issuing petrol coupons before going to teachers' training college in London. Spent thirty-five years teaching woodwork, metalwork and technical subjects in Cambridge schools, retiring in 1985. A member of The College of Handicrafts, obtained a BA from the Open University. Lived in Sheringham, Norfolk.

Moody, Dave   After service returned to Lockerley near Romsey, Hampshire, to work on his father's smallholding. Became an engine driver at a Royal Naval depot then went to work for Briggs Motor Bodies at Swaythling, Southampton, for thirty-eight years. Lives Southampton.

Munford, Alan   Farm worker from King's Cliffe near Peterborough, Northants, worked as a railway porter in London after release, then for a weighing machine manufacturer. At twenty-eight became a Baptist minister in Wimbledon, where he stayed until he retired. Lives Urchfont, 5 miles from Devizes.

Owen, Denys   Audit clerk in a Liverpool chartered accountants, became an articled clerk in Rhyl, then worked for other accountancy firms in Liverpool,

Ludlow, Rhyl for a second time, and Colwyn Bay. Went into partnership in a firm that expanded into St Asaph in north Wales, where he remained until he sold up at sixty-three. Still living there.

Pearce, Morry   After service returned to Monk Sherborne, Hampshire, to his job in the bakery and grocery store that had been his father's but was now taken over. Later went to Lancing Bagnall forklift truck manufacturer and worked in the tool room for thirty-five years. Lives Baughurst, Hants.

Potts, John   Post Office clerk from Wellingborough, Northamptonshire, spent lifetime in the Inland Revenue in different parts of the country, including twenty years in Sheffield. Retired as principal recovery officer for South East England. Home in Chesterfield, Derbyshire.

Rainbow, Peter   Office boy from Yatton, Somerset, who after leaving mining went to work for London, Midland and Scottish Railways before joining ICI. Spent forty-five years as an explosives engineer in Rhodesia and South Africa. Lives in his home village.

Ralston, George   Returned to the steelworks in Corby, Northants, in which he was employed when conscripted, working first in general stores and then in the engineering shop. From the age of fifty-nine, when the steelworks closed, was a cost clerk in a local company making lifts.

Reekie, David   Accounts clerk in a central London publishing company, did not return to his home in Deptford after release, but stayed in Yorkshire – as a clerical officer in the Leeds House Coal Office, which allocated supplies. Entered the brewery trade, working in the Midlands and then back in Yorkshire, at Ilkley. Became tenant of The Willow Tree Inn at Leeming, then oversaw the growth of a Northallerton packaging business. Retired at seventy-three and enrolled on a Leeds University creative writing course. Lives Bedale, where he is a founder member of the Bedale Writers.

Robinson, Phil   Grammar school boy from Wallasey, Wirral, who stayed in the coal industry, for thirty-nine years a mine surveyor in south Lancs and north Wales, where he lives, in Wrexham.

Roland, David   Training to be a cutter in a children's clothing business in Hackney, he came out to spend the next twenty-five years as a London black cab driver. His last fifteen years were as a salesman in menswear in the City. Lives Pinner, Middlesex.

Sadler, Ken   Co-operative Wholesale Society salesman in Newcastle on call-up, returned to the CWS, working in the drapery department. After winning a company management scholarship, moved to Manchester where he retired as senior buyer controller in ladies' fashion. Lives Sale, Cheshire.

Selby, Deryck   Mechanical engineering apprentice from Walthamstow. After finishing his professional qualifications worked for Standard Telephones and Cables, and other telecommunication companies. Lives Buckhurst Hill, Essex.

Simons, Victor   German refugee who made mining his career, joining the Coal Board research department in 1953 after getting a PhD at Sheffield University. Became area method study engineer in the North East Division based in Barnsley. In 1969 went to the Gas Council and retired as assistant manager of development in London. Home, Hampstead.

Spencer, Roger   Railway clerk from Richmond, Yorkshire, went back to the railways. Ended a forty-year career as a works study practitioner. Lived in Lytham St Anne, Lancs.

Squibb, John   Printer in his father's business in Weymouth, spent his career with the *Dorset Evening Echo* as a compositor, proof reader and VDU operator. Lives Weymouth.

Taylor, Reg   Police force civilian worker in Bradford, Taylor became a freelance newspaper photographer and retired as deputy pictures editor of the *Newcastle Evening Chronicle and Journal*. Lives Cullercoats near Tynemouth on the Tyne and Wear coast.

Taylor, Warwick   Junior clerk from Harrow, Middlesex, went to work for British Overseas Airways (later British Airways) at Heathrow as a station assistant, then in stores, then in public relations, ending a thirty-two-year career as a duty manager. Spent another eight years with Customs and Excise, also at Heathrow. Awarded an MBE in the millennium New Year's Honours. Lives Poundbury, Dorset.

Thomas, Les   Cashier at a London fountain pen company, on release went to work for an American airline. When it was taken over by another airline, moved to an Australian bank, where he retired as manager of the international department. Lives Rayleigh, Essex.

Thompson, Derek   Diamond tool apprentice from Salford, he was unable to find a job on release and joined the RAF, working in pay accounts in Britain and Germany for twelve years. Coming out as a corporal, had a spell as a Civil Service clerk, then worked in the county treasurer's department in Salford, retiring as a senior officer. Lives Blackpool.

Tyers, Ken Labourer from Tow Law, in the Wear Valley, County Durham,. on call-up, worked in heavy industry including brickworks and steelworks. Lives Consett in his home county.

Walker, Syd On release, worked in the costing office of a Birmingham safety glass company before studying arts at Ruskin Hall, Bourneville, obtaining a National Diploma at Birmingham college of arts and crafts, and then becoming qualified to teach, which he did for three years. In 1957 he opened his own pottery studio and gallery in Montrose, on the east coast of Scotland. Awarded an MBE in 1998 for his contribution to art in the community.

Warren, Derrick Returned to Battersea power station ('to burn coal in a big way') for several years, before office jobs with an electrical company and a radio and TV outlet, before working for a supplier of carbon on the technical side. Later a jobbing printer and owned his own business. Home in Swansea.

Weiss, Meir Farmed on release until allowed to emigrate to the new state of Israel in December 1948, where he spent twenty-five years on kibbutzim, then worked as a Ministry of Agriculture admin officer. Returned to England and ran his own soft fruit farm at Mortimer, near Reading for the next twenty years. Still lives nearby.

White, Dan Diesel mechanic from Hastings, became a policeman in West Sussex, first in Lancing, then Westbourne and after making sergeant in Billingshurst. He finished his service in Worthing, where he still lives.

Whittle, Donald From Eccles, Manchester. Returned to Magdalen college, Oxford, to finish degree, then to Cambridge for a degree in theology. Became a Methodist minister like his father. Worked in Gravesend and Leeds, then for nine years at a Methodist boarding school in Harrogate as chaplain and teacher. Lectured in Cambridge and Bath and did consultancy work in Wiltshire. Home, Bishops Castle, Shropshire.

Wilson, Les Did not return to Sainsbury's in Brighton, but joined the local bus company where he worked in the body shop until he retired. Still living locally.

Yates, Phil Office boy in a Winchester solicitor's, returned to the firm, qualified as a solicitor's clerk, and took early retirement as senior legal executive in 1983 to work as marketing officer at the city's Theatre Royal. Still working with the theatre, often as duty manager. Lives nearby.

Notes

Chapter One

1. In 1938, a Schedule of Reserved Occupations was drawn up, exempting certain key skills from conscription; the government was determined not to repeat the mistake of the First World War, when the over-eager recruitment of men into the military left war production dangerously short of workers. The scheme was complicated, covering 5 million men in a vast range of jobs, including miners – but at this stage only those underground, not surface workers; they were given reserved status after the war started. Railway and dock workers, agricultural workers, doctors and schoolteachers were in reserved occupations as were those in engineering, the industry with the highest number of exemptions. Ages of reservation varied between industries and could change, the government frequently reviewing the situation as the need for men to join the armed forces grew greater.

2. *Through the Dark Night* (Gollancz, 1941).

3. The weekly minimums were £4.3s underground and £3.18s on the surface. At the outbreak of war miners were earning an average of £2 13s 9d and stood eighty-first in the list of industrial wage rates. This second increase of the war, to an average of £5 2s 5d, only raised them from fifty-fourth to twenty-third.

4. The assumption from the way in which Bevin described the ballot is that, other than on dates when two draws were held, a consistent tenth of registering men were diverted into coalmining. The table below, relating only to the Midlands region, not only shows this wasn't so but that on some occasions numbers were drawn on a regional basis.

    The fifteen single-figure draws in 1944 and the one in 1945 were probably national, taking one man in ten; the double draws (in December 1943; and January, October and November 1944) were also likely to have been across the board, taking one in five.

    But the nine double-digit ballots (in March and May 1944, and all but January's in 1945) seem to indicate that only one man in a hundred was picked. As it is statistically impossible that Bevin could have conscripted the numbers he did if the nine double-digit draws were nationwide, then these must have applied regionally.

Whether this interpretation is correct – presumably to control intakes, dependent on the capacity of the training centres and perhaps an 'evening out' across the country – or whether other regions also had regional draws from time to time, can't be answered. No coalmining records other than those of the Midlands have survived. A lot were lost in the transfer of the industry from the colliery owners to the Coal Board on nationalisation in 1947; and in the 1950s the Public Records Office, now the National Archives, destroyed the bulk of what it held, for reasons of space, keeping only the Midlands records as being representative.

Dates	Numbers Drawn
1943	
14 December	0 & 9
1944	
15 January	6 & 9
29 January	2 & 5
16 February	7
1 March	2
16 March	64
31 March	9
15 April	8
8 May	13
20 May	8
3 June	5
20 June	1
10 July	1
24 July	9
5 August	3
19 August	3
2 September	0
16 September	4
7 October	8 & 9
18 October	1 & 2
1 November	0 & 2
22 November	4 & 6
6 December	8
20 December	6
1945	
3 January	0

17 January	30
5 February	37
19 February	74
5 March	23
23 March	40
6 April	33
23 April	11

Note: One other ballot was held, on 7 May 1945 (in the Midlands it drew the double-digit 36). Victory in Europe brought its cancellation and the end of the Bevin ballot scheme.

5. Anthony Shaffer died in 2001, the year in which Picador published his memoir, *So What Did You Expect?* He remained a miner until 1948; his brother developed an ulcer and was released after two years. Anthony is best known for his thriller *Sleuth*, later filmed with Sir Laurence Olivier and Michael Caine. Sir Peter is best known for *Equus* and *Amadeus*, also filmed.

6. The change in the public's attitude to women working was astonishing. Before the war, a single woman who took a job she didn't need was frowned on for taking it from someone who did need it. The Civil Service, the BBC, nursing and teaching were among careers from which a woman had no choice but to resign when she married.

7. Not much had changed thirty years on. In 1976, the year he died, Sir William Lawther, former Labour MP and wartime president of the Federation, recalled parents in tears seeking help to get their sons exempted. 'You would have thought the boys were being sent to Dartmoor in chains,' he said.

8. Throughout the war male students in science, engineering, medicine and dentistry were able to finish their courses, unless they failed examinations. For arts students it was different. Until 1941 it was possible for them to complete two years at university provided they showed 'exceptional promise as potential officers or intellectual ability above the average, or evidence of a balanced combination of the two'. From March 1942, however, two years was cut to one and from the December, all arts students were conscripted at eighteen. In the period up to then, universities accepted seventeen-year-olds so that they could complete a year before call-up.

9. Whether the South Shields apprentice went down the pits I've been unable to trace. Surely he did: the government could hardly have given in on the issue.

On one occasion during the post-war period of national service, the government of the day resorted to the same tactic. In June 1955 an unofficial strike of merchant seamen – a group exempt from military service – was preventing liners in Southampton and Liverpool from sailing. Two hundred men of conscription age were sent notices of call-up. They immediately re-registered with the shipping pool.

Chapter Two

1. The nine English training centre collieries, from north to south: Lamb, at Cramlington, Northumberland; Horden (with nearby Easington colliery as an auxiliary establishment), at Horden, and Morrison, at Annfield Plain, County Durham; Prince of Wales, at Pontefract, Birley East, at Woodhouse, and Askern Main, at Doncaster, Yorkshire; Newtown, at Swinton, Lancashire; Creswell, at Creswell, Derbyshire; and Haunchwood, at Nuneaton, Warwickshire.

   Additionally, the Kemball colliery in Stoke-on-Trent, which in 1942 had established a training centre for boys in North Staffordshire coming into the mining industry, became part of the government network. In Kent, the Chislet colliery took on training for the four pits in the county.

   The Scottish centre was at Muircockhall, near Dunfermline in Fife; the Welsh at Oakdale colliery at Blackwood near Newport in Monmouthshire.

2. Run on a non-profit basis by local authorities at the government's request, around 2,000 British Restaurants were set up during the war in schools, church halls and any other available premises to offer the public a cheap meal (three courses: minced beef with carrots and parsnips a typical main dish, 9d) for which they didn't have to surrender coupons from their ration book. The government encouraged people to use them once a week as a way to supplement their rations. British Restaurants were very basic but popular, with long queues.

3. Use of the Coventry Sally continued well into 1944. On 31 March George Griffiths, the Labour MP for Hemsworth, raised the conditions there in the House. According to the *Daily Worker*, he said, there were not enough baths, the food was poor, men, often mining trainees, went to work without breakfast – and when they went home for a weekend, tramps were allowed to sleep in their beds. Bevin replied that there were twelve baths and five more were being added; the food was good; and the beds of men away for a weekend were reserved for them. It was true that a night watchman, since dismissed, had on one occasion allowed the beds to be used by other workmen. Bevin said the *Daily Worker* article 'is a tissue of lies, but that is the Communist method of supporting the war'.

4. In relation to what perhaps was a more comparable group – the engineering apprentices – the conscripts were also poorly off. A trainee engineer of eighteen had his travelling expenses to and from his training centre paid, and he got his midday meals free. He wasn't paid a wage, instead receiving an allowance of 10s 6d, with a proficiency bonus, and then a second, later in his training. Crucially his biggest single expense, his lodgings, was paid for.

5. The demand for a subsistence allowance could never have been granted because it would have raised the question of equal treatment for the tens of

thousands of men and women directed into other industrial jobs away from their homes.

6. Allegations that men went into the pits in the First World War to avoid having to fight became a bitter national issue, as this extract from *The Colliery Guardian* of 17 March 1916 indicates:

'Some rather wild statements have been made as to the number of men of military age who have entered the pits with the object of evading military service. These allegations have been emphatically denied by Mr Smillie, the president of the Miners' Federation, who says that a large proportion of the 160,000 fresh men who have entered the pits since the outbreak of war are youths and old men. On the other hand, Lord Derby, in the Hous of Lords on Wednesday, definitely declared that there were a great many men who had gone to work in the mines in order to avoid military service, and he hoped that these men would be taken out and made to join the Army.'

7. Ten thousand unaccompanied children, most of them Jews, were brought out of Germany, Austria and Czechoslovakia prior to the invasion of Poland. A plaque marking the fiftieth anniversary of the Kindertransports was unveiled in Harwich, where the first boat arrived from the Hook of Holland, on 14 June 1989. On 16 September 2003, a commemorative statue was unveiled at Liverpool Street station, the London terminus where the second leg of the young refugees' journey ended.

8. *My Farce from My Elbow* (Star Books, 1977). Rix – Lord Rix – paid only a flying visit to mining: he was gone in three months. After training he was assigned to his training colliery, Askern Main, and worked underground. But after a medical examination it was concluded that depth would be as much a problem for him as altitude. He was returned to the air force, serving at RAF Halton in Buckinghamshire as an instructor, later taking charge of medical staff postings.

9. The first conscientious objector that the Peace Pledge Union (a prime source on conscientious objection) knows to have been given the choice of the mines as well as the normal alternatives of the land, hospitals or the fire service, was the well-known writer on criminology Tony Parker (1923–96), in early 1942. By August 1943, at least one division of the Appellate Tribunal routinely allowed coalmining as a condition of military exemption. There is no record of how many men took that route.

On 17 September 1943, the Midlands tribunal gave Maurice Bailey of Birmingham coalmining as his exemption. He objected on the grounds that the major need for coal was the war factories. After a furore in the local press and the intervention of the Anglican Bishop of Birmingham, Ernest Barnes (a noted pacifist), the Appellate Tribunal upheld Bailey's appeal and he served with the Friends Ambulance Unit.

10. *Muck, Silk and Socialism: Recollections of a Left-wing Queen's Counsel* (Paper Publishing, 2002).

11. Several thousand Indians came to England during the war to work in the factories as 'Bevin trainees', but were informally referred to as 'Bevin Boys', including in the Indian parliament; the term was later applied to entrants to the Foreign Office when Bevin was Foreign Secretary (July 1945–March 1951) and entered the language widely in a facetious way.

12. *Bevin Boy – A Reluctant Miner* (Athena Press, 2004)

13. According to several Bevin Boys who were at Newtown training centre at Swinton, near Manchester in May 1944, a curious rumour that went the rounds was that the PTIs were professional toughs who used to tour with one of the local fairs.

14. When Geoffrey Mockford and his intake were waiting to sign on at Chesterfield labour exchange before commencing training at Creswell colliery, 'we started to compare our National Registration numbers. Every number from nought to nine was represented. I never found an explanation for this.'

    This could be taken as another indication that some numbers were drawn on a regional basis (Note 4, Chapter 1). However, if there were ex-servicemen and optants in the intake, their numbers would have been different. As Mockford had appealed and was therefore held back for more than a month, he was reporting with a later intake. His number would have been different from these, as it would have been from any who appealed against the ballot drawn on different dates.

    In Frank Pratt's case, while his last number could conceivably be regional, the likelihood was that a mistake was made. A few Bevin Boys who found themselves in a similar position complained to the Ministry of Labour, which admitted the error – and told them to get on with it.

15. Out of the total of 45,859 men from all sources, 1,800 were failed at the training centres, their files stamped UNSUITABLE.

16. In the interests of accuracy – not all collieries had slag heaps. A few on the coast emptied their rubble into the sea via overhead cables.

Chapter Three

1. *The Bevin Boy* (ISIS Publishing, 1995).

2. If suitable for conversion to coke, coal up to 1½in was transferred to a colliery's coke ovens, which provided the site with heating. Coal over 8in was usually cleaned by hand. Processed coal was taken to the massive storage area to await loading and despatch by rail, or in some areas by canal barge.

3. *As It Happens* (Barrie and Jenkins, 1974).

4. Until the middle of the nineteenth century, colliers widely bound themselves to a pit owner for a period of a year without the power to terminate the contract.

It was known as the bond. The practice died out, but in Wales the word became a colloquialism for the cage.

5. The overseer or foreman came between the underground manager and the deputy (in some pits called the fireman), whose responsibility was the safety of a working area known as a district. Pits could have numerous districts.

6. The base chamber of a carbide lamp was filled with carbide chips, like irregular lumps of sugar, on to which water from a chamber above was dripped, controlled by an external stopcock. The resultant fumes (acetylene gas) were piped to a tiny jet mounted in front of a polished reflector the size and shape of a saucer. The lamp had a flint lighter and could therefore be relit. Miners usually carried extra carbide in an old cocoa tin.

7. Sulphurated hydrogen was produced by the decomposition of iron pyrites, the golden specks sometimes found in coal, in wet workings.

8. The onsetter was in charge of winding operations underground, responsible for loading and unloading the tubs from the cages and instructing the winding house.

9. Tom McGuinness, known as the miners' artist, received an entirely different response. A Bevin Boy working in Fishburn colliery in his native County Durham, he chalked images of the pit on the side of the tubs but his deputy encouraged him to take art classes. McGuinness, who died in 2005, left the pit in 1948 but returned, spending his working life in mining.

10. Chalking on tubs was a crude way of informing the checkweigh man who had hewed the coal; chalk marks were often partially erased in the handling of the tubs – and miners also chalked on tubs to pass messages, whether it was to do with major events in the war or the winner of the 2.30 at a race meeting. A better method used in some pits was to daub different coloured whitewash on the topmost coal in a tub. A better method still: in other pits miners carried a bundle of tokens – brass, iron, lead, even leather – stamped with their number; one was hung inside each tub they filled.

11. Main and tail was a single-track, haulage system operated by an engine with two drums, each with a separate rope.

12. In 1934 only 48 per cent of British collieries were mechanised, compared with nearly 100 per cent in Belgium and the Ruhr. By the outbreak of war, the percentage had risen to nearly two-thirds.

   The mechanised cutter was essentially a steel box 5ft long, 3 wide and 2 high. Operated by compressed air, it moved along the whole face with a chainsaw running from a long arm, tearing out a 4in-thick cut 5ft into the coal at about half the height of the seam. Then the shot firers drilled holes along the top of the face at regular intervals under the control of the shot firer (fireman). Many pits still used gunpowder, rammed in with damp clay or the dust from the drillings wrapped in rolls of newspaper; more advanced pits used steel shells, 4ft in length and 2in in diameter, containing solid carbon dioxide at a pressure of many thousands of pounds to the square inch.

Most shot firing was generally done on the night shift, bringing down some coal but more importantly fracturing the face for the collieries to start work next day, when it was also carried out as required.

13. A stint (or stent) was the width of coalface, usually 3 to 5yds, which the collier was expected to work in a shift.

14. *The Bevin Boy* (ISIS Publishing, 1995).

15. The Mines Act of 1847 barred children from much of the heavy underground labour in British mines, which was when ponies came from Shetland and Galloway, Exmoor and Dartmoor. No pony under four was allowed down the pits; collieries had breed programmes but supply never met demand and ponies were imported from Iceland, Scandinavia, Russia, Belgium and even America. In 1919 the lowest price an animal fetched was £26, the highest, £45. In 1943 the lowest was £20, the highest £23.

There were over 70,000 ponies and horses in the pits before the First World War; when the Second ended, 60 per cent of the 23,000 animals that remained were in Durham and Northumberland, some pits having as many as 200. The majority of cob horses were in south Wales. In 1945 a writer to *The Times* hoped for 'a new era' in mining, where 'the pit pony is the symbol of a bygone, ugly age'. Yet the 1957 Pit Ponies' Protection Society's annual report noted that 12,000 animals were still underground. Astonishingly the last, at Ellington colliery in Northumberland, came up only in 1993.

16. A closing – the roof or sides of a roadway coming down or in due to strata pressure.

17. *Muck, Silk and Socialism* (Paper Publishing, 2002).

18. In a pit explosion, coal dust is lifted into the air. Each particle absorbs oxygen from the atmosphere allowing the flame to transmit from particle to particle. Stone dust, being inert and inflammable, breaks the chain. Collieries ground whatever stone came out of their pits, but limestone was favoured because, being soft, it ground very finely. The safe ratio of stone dust to coal dust was generally regarded as about two to one.

19. The high acid content of mine water was due to the weathering of iron pyrites.

Chapter Four

1. Unlike other food stuffs, bread wasn't rationed during the war (though it became so in 1946 – see Chapter 8). From 1942, the Federation of Bakers was formed to organise the production and distribution of the National Loaf, which was introduced due to the shortage of white flour. Almost all pre-war bread was white and many people considered the National Loaf to be coarse

and unappetising because of its high bran content. Roughly speaking it was the equivalent of today's brown bread.

While rationing went up and down a little depending on supply, the average adult was entitled to a weekly allowance of: 4oz bacon or ham; 2½ pints of milk; 2oz cheese; 2oz butter; 2oz margarine; 2oz fat or lard; 2oz tea; 2oz jam; 3oz sugar; 3oz sweets; 2lb onions (rationed between 1942–44). Meat was rationed by price: 1s.2d (roughly 1lb 3oz). Eggs were scarce from 1940 following cuts in imports and the slaughter of millions of hens to save feeding stuffs. The ration was one a fortnight, but there were long periods with none at all. From June 1942, by which the Battle of the Atlantic was largely won, dried egg from American became available – a tin every eight weeks for an adult, the equivalent of twelve fresh eggs. 'Shell eggs are five-sixth water: why import water?' ran a government slogan. In addition to the weekly ration, everyone had four points for cereals, biscuits, dried fruit and tinned vegetables, which could be saved.

2. Up to the First World War, scarves in Wales were an indication of miners' status. Only the colliers – the elite – were allowed to wear theirs wound around the neck. Most others could wear them but only crossed over the chest. Some jobs were considered too lowly for those who did them to be allowed to wear a scarf, including most surface workers, irrespective of the weather.

3. In Wales, some elderly miners showed John Wiffen their shins, which were pitted and scarred as the result of clog fights, 'a type of sport in which those taking part faced each other in pairs, put their hands on their opponent's shoulders and at a given signal kicked hell out of each other's shins. The winner was the man who could take most.'

4. *Bevin Boy* (George Allen & Unwin, 1947).

5. Some Bevin Boys went to pits in north Wales, but almost all served in the eastern part of the coalfield. The decision not to send them to the pits in the west, where the manpower situation was most critical, was because most miners there spoke only Welsh and language difficulties underground would have been dangerous.

6. 'Kenner' was another term Midlands term, also used in places in the North East, where 'coat it' was only less popular than 'loose it'.

7. *The Bevin Boy* (ISIS Publishing, 1995).

8. The Miners' Welfare Fund spent £6.5 million on pithead baths during the war. The Fund also opened fifteen convalescent homes (£3 million), established seven medical rehabilitation centres, and spent £1.25 million on surgical treatment and appliances.

Formed in 1920, the Fund got its money from a levy of a penny a ton paid by the colliery companies on the coal produced and a levy of a shilling in the pound on the royalties paid to those who owned the land from under which the coal was dug.

9. *Bevin Boy* (George Allen & Unwin, 1947).
10. Soap rationing was severe, except for shaving soap – men needed to shave. People soon caught on to the idea that small batches of hand washing could be done with shaving soap, especially if several sticks were rolled together. Shops refused to sell shaving soap to women, but it wasn't unusual to see a man buy two or three sticks at a time.
11. *The Bevin Boy* (ISIS Publishing, 1995).
12. At the annual conference of the Co-operative Party in April 1944, the need for additional food for miners was urged by several delegates. Miners' wives, one delegate asserted, didn't want the canteens at the mines, they wanted to be able to put a good solid meal before their men when they returned home. The delegate from Nottingham said that the canteens in his district served first-class meals but fewer than 50 per cent of the men turned up to eat them. A resolution was adopted urging the Ministry of Food to grant miners extra rations of meat and bacon.

Chapter Five

1. *Muck, Silk and Socialism* (Paper Publishing, 2002).
2. The method used to control an underground fire was to seal off the seam with a brick wall, to prevent air getting to it, and let it burn out as oxygen was exhausted. Large areas of coal were lost in this way: the seam wasn't reopened.
3. It's difficult to make a clear case from the figures. After the First World War, the death rate, which during it was 1.33 per thousand men, declined to 1.05 in 1938, when 858 were killed, but from a larger workforce than in 1942 when the death rate rose to 1.14. After 1942 the rate gradually reduced, standing at 0.83 in 1945, when 550 men lost their lives. By 1947, the last full year of the Bevin Boy scheme, the death rate was 0.44.

   The seriously injured table (given not by numbers of men but by man-shifts worked) in 1938 was 1.52. This dropped to 1.49 in 1939, rose to 1.52 in 1940, but thereafter gradually reduced, standing at 1.31 at the war's end.
4. The Mass Observation organisation was funded in 1937 by two Cambridge drop-outs to study everyday life in Britain. Voluntary observers went to every kind of public event, spoke to people at work and in the street, and recorded behaviour and overheard conversations.

   The organisation claimed to have 2,000 observers, which was almost certainly a figure inflated to give it authority, and it lacked funds to mount surveys with adequate samples. But if not rigorous, its findings, mostly published in pamphlets, excited a largely youthful readership and its distinctly left of centre stance probably had some influence on the Conservative Party's landslide defeat in 1945.

Mass Observation did have a nucleus of paid investigators of whom the late Eric Gulliver was one. In his time down the pit, Gulliver sent back hundreds of reports and views gathered in interviews or simply overheard. All his material is in the Mass Observation archive at the University of Sussex, which he helped set up.

5. Safety lamps went through several improvements by other men after Sir Humphrey Davy in 1815 used a wire gauze to protect the flame (which prevented it from coming into contact with volatile gases), but his name remained as the generic descriptor.

Three gases presented dangers belowground: firedamp, blackdamp and afterdamp.

Firedamp was an inflammable mixture of methane (which occurs naturally in coal) and air. Lighter than air, colourless and tasteless, firedamp lodged near the roof and at around 2 per cent was liable to explode if it came into contact with a naked flame, usually from a defective lamp gauze, an opened lamp, a 'blown-out' shot, or sparks from the fall of hard stone from the roof or from a pick striking stone or iron pyrites.

In the presence of firedamp, a cone of green-blue flame hovered above the safety lamp's wick flame.

Blackdamp, a mixture of nitrogen and carbon dioxide (CO_2), usually found in old workings or badly ventilated headings (new roadways still 'blind' at one end), could in large amounts choke a man – at 15 per cent it made breathing difficult. It was tested for by lifting the safety lamp up and down, the telltale flame rising and falling with the movement.

Afterdamp (carbon monoxide, CO), a product of the incomplete combustion of carbon, found after a fire or explosion – and once the cause of more deaths than either – was the most poisonous: as little as ½ per cent caused giddiness and 1 per cent, death.

During rescue operations after a disaster, rescue team members had breathing apparatus, but as the amount of their air was limited, they didn't use it until absolutely necessary. To avoid delays that stopping to test for gas with a safety lamp would have meant, they carried two or three canaries in separate cages. In the presence of gas, the birds would become distressed and fall off the perch. The last canaries were phased out of British pits only in 1986, replaced by electronic detectors.

6. Yet other collieries had a stock of new lamps and at least a few expected Bevin Boys to buy theirs. As with picks and shovels, most refused. In such cases it wasn't unknown for a colliery to issue a Bevin Boy with a safety lamp, which threw insufficient light for him to work by so that, however reluctantly, he eventually preferred to pay up.

7. Pratt went back to work in the factory he'd been in before call-up. When his father, a twenty-six-year regular, was released from the army and came home,

Pratt was called up again and served as a corporal in his father's regiment, the Dorsets.

8. *The Bevin Boy* (ISIS Publishing, 1995).

9. In April 1945 the Society reported that in one drift mine, every one of twenty-two ponies was into his second shift in twenty-four hours and one pony was in his fourth shift in five days. Three had been worked to death in three months.

10. A cotrell (also called a cockle in Durham) was a spring clip shackle-pin that attached the limber [shaft] to the tub.

11. A device not unlike the plates at the exit of many car parks today, which was set between the rails at the head of an incline. It allowed the ascending tubs to pass over, then lifted to form a barrier in case any broke loose from the haulage rope. In most pits a trailing piece of metal like a crowbar, known as a drag, was attached to the last tub in a set; if there was a runback, the drag derailed it.

12. In the period of the war, 6,000 men in south Wales alone suffered from silicosis or pneumoconiosis. In *Bevin Boy*, Derek Agnew quoted extracts from a *Daily Mirror* article, 'Murder by Dust', written by John Hogan:

'I met eleven miners on the barren mountain overlooking Ferndale, in the unhappy Rhondda Valley. They had climbed the twisting paths to breathe the pure air. They made a pathetic job of it.

They coughed for minutes and twisted with pain, overcome by the keen freshness of the mountain breeze. Each of them explained he had "bronchitis".

They sounded as certain as men trying to convince themselves a goldfish could swallow a horse.

"They're not fooled," my companion told me. "They know the dust has got them, but they can't help hoping . . ."

"You can take out the lung of a miner who's died of lung disease and drop it," ex-miner Jim Griffiths, Minister of National Insurance in the post-war Labour government, once explained graphically, "and it hits the floor like a stone. And if you try to cut it, it's like trying to slice a piece of thick, tough leather."'

Nystagmus was once the most prevalent of all industrial diseases and was found only among miners – 2,262 men in mining were certified with the disease in the first six months of 1945. The disease, caused by working in the poor light of safety lamps ('stag' wasn't so common in naked-flame districts) affected vision, its main symptom being uncontrolled, jerky side-to-side movement of the eyes. Usually it took twenty or thirty years to appear.

13. In 1982 Australian researchers Barry Marshall and Robin Warren discovered a bacterium later named *Helicobacter pylori* (H. pylori) that they speculated caused gastritis and peptic ulcers – heresy at the time, medical belief being that bacteria couldn't live in the stomach's acid environment.

Marshall proved his point by swallowing a solution of the bacteria to make himself ill and then cured himself with antibiotics.

14. *Polymyelitis*, or infantile paralysis as it was inaccurately called in the 1940s, affected eighteen people in 100,000. About half made a complete recovery, 20 per cent were left with a slight handicap and 25 per cent were crippled. In 5 per cent of cases, polio was fatal. The virus was largely eliminated in the early 1950s with the introduction of the Salk vaccine.

15. Extrapolating from statistics can give questionable conclusions, but the death rate per thousand men (Note 3 this chapter) in 1942 was 1.14, declining to 0.44 in 1947. Taking the mean of the two figures (0.79) and applying it to 48,000 Bevin Boys could indicate as many as sixty-four might have been killed; even 0.44 indicates twenty-one.

Chapter Six

1. *So What Did You Expect?* (Picador, 2001).
2. *Bevin Boy* (George Allen & Unwin, 1947).
3. Churchill began Parliamentary life in 1900 as a Conservative for Oldham, crossed the floor to the Liberals, lost his seat but got another in Dundee, which he held until 1922, after which he was out of the Commons for two years. In 1924 he was elected in Epping and subsequently rejoined the Conservatives. In recognition that the constituency had changed from a country seat to a populous borough, Epping became Woodford for the 1945 election in which the Conservatives were defeated. Churchill held the seat until he stood down in the 1963 election.
4. Mansfield, Rufford, Clipstone, Thoresby.
5. *Bevin Boy* (George Allen & Unwin, 1947).
6. Rab Butler was briefly the second Minister of Labour and National Service (May–July 1945) after Bevin became Foreign Secretary and was followed by Isaacs, who held the post for five years. Four more ministers held the joint post before national service was abandoned in 1960.
7. At the peak of British coalmining, during the First World War, there were over 2,500 pits. At the beginning of the Second, there were 1,900, but around 250 fewer by the end as seams were worked out or became unprofitable.
8. Life is nothing if not inconsistent: the union lodge in Cefn Coed colliery near Neath actually refused to allow Bevin Boys to join until forced to do so by the Federation.

Chapter Seven

1. Conscripts went to the coalfields in the following numbers: 1. Yorkshire, 5,027; 2. Midlands (Nottinghamshire, Derbyshire, Leicestershire, Staffordshire, Warwickshire, Shropshire) 4,474; 3. County Durham, 4,115; Wales (plus the

Forest of Dean and Bristol) 2,586; Scotland, 1,305; 6. Lancashire, 1,302; 7. Northumberland and Cumberland, 1,165; 8. Kent, 260. Total: 20,234.

Numbers of ex-servicemen, optants and volunteers were: 1. Midlands, 6,219; Yorkshire, 4,986; 3. Durham, 2,663; 4. Lancashire, 2,560; 5. Wales, 2,315; 6. Scotland, 1,768; 7. Northumberland and Cumberland, 674; 8. Kent, 172. Total: 21,357.

2. Hostels ranged in size, the smallest having accommodation for 100 men, the largest for 600. Wales had five, Scotland nineteen and England twenty (twelve in the North East, three in the North West and five in the Midlands). Altogether the hostels' capacity was 15,270.

3. Whatever the government said about dried egg, there was no disguising the fact that it had a cardboardy taste. It was, however, excellent scrambled.

4. The blackout became the dim-out – officially half-lighting – at dusk on 17 September 1944. By December, lighting on trains was back to normal. The dim-out was abolished on 24 September 1945, except for a 5-mile coastal belt, where it remained in force until 10 May.

5. Oakdale's miners' institute was built in 1916, the funding lent by Tredegar Iron and Coal Company and paid back by the miners over the years. The institute is now part of St Fagans open air museum west of Cardiff on a site where over forty original buildings from Wales's past have been re-erected.

6. Nat Lofthouse OBE, who played thirty-three times for England from 1951, made his Bolton debut at fifteen and turned out for his club in 452 league matching, scoring a club record 256 goals.

One of the sons of a coal bagger, Lofthouse went down the Mossley pit, a forty-five-minute tram ride from home. During his service, every Tuesday and Thursday evening after a shift, he trained at Burnden Park, afterwards feeling 'like walking home on my knees'. If Bolton was playing away on a Saturday, the team coach picked him up at the colliery after the morning shift. 'I took a lot of ribbing from the miners,' he wrote in his autobiography, *Goals Galore* (Stanley Paul, 1954).

Becoming a Bevin Boy 'was the best thing that could have happened to me' Lofthouse has said in numerous interviews. In *Goals Galore* he wrote that pushing tubs 'made me fitter than ever I had been before. My body became firmer and harder. I learnt to take the hard knocks without feeling them. My legs became stronger and when I played football I felt I was shooting with greater power.'

Unfortunately Lofthouse declined to be interviewed for this book.

7. *The Bevin Boy* (ISIS Publishing, 1995).

8. *The Bevin Boy* (ISIS Publishing, 1995).

9. All the leading British orchestras played for ENSA – the Entertainments National Service Association, set up on the outbreak of war – at some time; between October 1943 and May 1946 nearly 400 full-scale symphony concerts were

performed all over the country. Serious music and drama were a sideline for ENSA but the main activity of CEMA, the Council for the Encouragement of Music and the Arts, which also took opera and ballet to audiences that had never seen them before.

Chapter Eight

1. The first serious strike of the war, in 1941, involved not miners but engineering apprentices, first on Clydeside and then in Coventry, Lancashire and London in pursuit of a pay increase in line with time-served engineers. The Lancashire apprentices only returned when they became the first group of workers to be prosecuted under the government's emergency anti-strike legislation.

Almost all strikes were in support of wage demands; the majority of the remainder to protect against changes in working conditions – in the House in September 1943 Bevin talked about 'some strikes [being] deliberately provoked for ulterior motives by employers – to create the conditions to change workers' contracts'. British wartime industrial relations are outside the purpose of this book, but the following gives an indication that the miners weren't the only workers to go on strike:

April 1942, the Ministry of Labour attempted to direct 400 men from the Royal Ordnance Factory in Nottingham to private companies, which involved a reduction in rates of pay; the workforce occupied the factory. In the September Tyneside shipyards struck when management changed working practices without consultation. On Tyneside at the beginning of 1943, workers at the Neptune ship repair yard came out for six weeks after the refusal of five men at their firm to join the Amalgamated Engineering union. Massive support from workers and other firms and trades forced the employers to concede a closed shop. In the September, Vickers' shipyard at Barrow was shut down by 9,000 workers for eighteen days over a pay award. Other stoppages of the year involved 12,000 bus drivers and conductors and dockers in Liverpool and Birkenhead. Also in 1943: skilled female workers at the Rolls Royce aircraft factory in Glasgow struck over equal pay – something that has been agreed in 1940 and not implemented. The women won a partial victory: their pay was upped, not to the level of a male skilled worker but to that of a male semi-skilled worker. The major stoppage of 1944 brought out nearly 30,000 men in Belfast shipyards and Northern Ireland aircraft factories in protest against the imprisonment by the Northern Ireland government of five shop stewards.

Sometimes strikes were nothing to do with pay and conditions. In 1944, 2,000 employees engage in a sit-down strike at Fairey Aviation aircraft factory 'in the north-west of England' over an alleged indiscreet remark by a charge-hand to a worker. Employees at another Fairey factory 3 miles away took similar action in sympathy. And to Bevin's embarrassment the first union protests to his amendment to the defence regulations after the Porter pay strike came from thousands of members of his own union, the GTWU, who tied up all bus traffic in parts of London and walked out of Manchester's municipal gasworks, leaving the city without supplies for a week.

2. *Muck, Silk and Socialism* (Paper Publishing, 2002).

3. In south Wales, a survey at the end of May 1944 found that out of 1,630 Bevin Boys only 6 per cent 'showed no interest in their work' but that 94 per cent were satisfactory, an indication that absenteeism was negligible.

4. In the first ten months to July 1944 in County Durham alone, 12,600 cases were deal with, an unknown number of them Bevin Boys.

5. The Land Army came into existence in June 1939. During their four weeks' training Land Girls received only 10s a week and their keep. In 1941 their pay was 28s, rising to 48s in 1944, keep excluded. Many were badly used by their employers, ill-fed and housed and they worked in the fields in all weathers, their boots and gaiters inadequate in wet, muddy fields; many didn't get a top coat for as long as eight months. Like Bevin Boys, a third of Land Girls lived in hostels, though these were much smaller.

6. The single white feather as a symbol of cowardice derived from eighteenth-century cockfighting and the belief that a cockerel sporting a white feather in its tail was cross-bred instead of pure-bred and therefore likely to be a poor fighter. At the beginning of the First World War, Penrose Fitzgerald, a retired admiral, announced in Folkstone that he'd formed a band of women to present a white feather – a fate 'far more terrible than anything they can meet in battle' – to young men 'found idling and loafing' instead of being in uniform. White feathers were given out all over the country but as the recipients were frequently men invalided from the trenches or otherwise engaged in the war effort, the practice was widely unpopular.

7. Platts-Mills failed to win the nomination of the King's Norton constituency in the south of Birmingham ('a Bevin Boy cut little ice compared with a captain in uniform'), but won the seemingly unwinnable seat of Finsbury in London. Always a thorn in the flesh of the Labour leadership, he was expelled from the party in 1948, remaining in Parliament as an independent until he lost his seat in the 1950 election. Made a QC in 1964, Platts-Mills defended the notorious gangster Charlie Richardson and one of the Kray twins, acted in the appeals of the Great Train Robbers, and appeared for the defence of trade unionists and alleged IRA terrorists in a number of trials. He was readmitted to the Labour Party in 1969. Throughout his long life (he died in 2001 at ninety-five) he continued to be associated with far-left causes – and always championed the miners.

8. All sporting events were curtailed in early 1947 because of the atrocious weather (see Chapter 9), though saving the energy used for floodlighting was

a factor. The temporary banning of greyhound racing, however, certainly owed something to the miners' preoccupation with it.

During 1946, Shinwell looked to see if there was a link between absenteeism and loss of output on one hand, and football and horse racing fixtures on the other. On 28 January, for instance, when a Cup-tie took place in Sheffield, he had the production at eight adjacent pits on that day compared with their averages. It was collectively down 4,761 tons down on 15,325. Absenteeism on that day averaged nearly 35 per cent – over 55 per cent at one colliery.

At Shinwell's instigation, a survey was carried out by the National Coal Board, formed on 1 January 1947 (again, see Chapter 9) to see 'whether evening racing (say after 7p.m.) would substantially interfere with production in their regions'. The three largest coalfields, Yorkshire, south Wales and Scotland, said yes. Others were less certain. The NCB subsequently conducted a survey in Staffordshire to examine the link between absenteeism and loss of output at different collieries 'close to and remote from dog tracks'. The result was inconclusive.

9. Britain came close to bread rationing during the dark days of 1941–42 but avoided having to introduce it. That it was deemed necessary in the first full year of peace was highly contentious. The Labour government justified the step as necessary to prevent famine in Europe, but the Conservatives doubted that substantial savings in wheat could be made and were backed by the bakers and the British Housewives' League.

The bread (and flour) ration scheme was more complicated than anything that had gone before and covered seven different categories of consumer. Essentially, the average person was entitled to 9oz a day; like the miners, other workers in heavy manual labour and agriculture, who also took sandwiches, got extra coupons.

10. As the government realised they would, other sections of heavy industry wanted an increased meat ration. Foundry workers, dockers and builders (who pointed out they didn't have canteens on building sites) made demands, which were resisted. But the allowance of meat to all category A industrial canteens was increased, from 2.7d to 4d a main meal served.

11. Dermatitis is an infectious disease. A man suffering from it would have been turned down by the army.

Chapter Nine

1. The £64 million state buy-out of the collieries was agreed by September 1944, but there were issues when the Nationalisation Bill came forward at the beginning of 1946, the colliery owners trying, unsuccessfully, to keep ancillary services connected with their businesses out of the pot. Around three-quarters of the 800 or so owners had been paid off by nationalisation.

The NCB took over 1,647 collieries (and ancillary services including coke ovens and wharves), more than 1 million acres of land, farms, villages and houses; 200,000 railway wagons, 23,000 pit ponies, and an undisclosed number of shunting engines and lorries.

The miners were particularly angry about the sums of money that went to the big landowners from under whose property coal was extracted. Before nationalisation such landowners were paid a royalty on every ton mined; now they received lump payments to buy out their rights. Four received a million pounds each, six over half a million.

One big landowner was the Church of England. 'It somehow sticks in my mind,' says Joe Geoghagen, a Catholic seminarist who became a Bevin Boy at the Dean and Chapter colliery near Darlington, 'that for every ton, 10 old pence were paid to the Church Commissioners in Durham. Church or not, this was greatly resented in the mining community.'

2. The Mines Department, forerunner of the Ministry of Fuel and Power, had prepared plans for the rationing of domestic coal at the beginning of the war. These were taken up by Hugh Dalton when he became President of the Board of Trade who, in April 1942, announced the decision to introduce fuel rationing from June, to be worked out by the renowned economist Sir William Beveridge. The public, having accepted food and clothes rationing, was prepared to accept coal being put on coupons.

Beveridge's scheme was ingenious. Coupons were to be handed over when coal or paraffin was delivered or collected, or the gas and electricity meters were read. Those over sixty-five would be able to exchange clothing coupons for extra fuel and the North would get a 30 per cent larger ration than the South.

Labour MPs, who hoped this was a first step toward nationalisation, were in favour, but opponents on the Tory benches, implacably wedded to private ownership, killed the introduction. The Ministry of Fuel fell back on the principle of 'voluntary rationing'.

The ordinary household allowance was a hundredweight of coal a month (in 1944, 3cwt in the South, 4cwt in the North) and because deliveries weren't always possible, to obtain it people walked long distances and waited hours. Often coke could be had from the local gasworks if it was fetched; women pushing prams or boxes on wheels were a common sight.

To supplement their meagre allowance, people gathered coal dust to mix with a little earth (or cement) and a little water, to make crude brickettes that spluttered fitfully without throwing much warmth. Many foraged for wood on bombsites. And, of course, conscientious citizens took their baths in 5in of water that was then shared by other members of the family.

Almost everyone lived in their kitchen, the only heated room in the house.

By the end of 1942 British homes used 4 million tons of coal less than in 1938. Another 4 million were saved in 1943 and 3 million more in 1944.

3. By the end of 1947 there were 1,635 stoppages, resulting in a loss of more than 1½ million tons.

4. At this stage, 2,788 Poles were working in the pits, the majority in the East Midlands; another 3,047 were in training but as yet unplaced.

5. Every miner was entitled to one parcel, which cost him 22s 6d. The parcels were not all the same but typically contained four 1lb tins of steak and kidney, two 10oz tins of peaches, 15oz sugar, 3oz of tea, 7oz of margarine, six packs of biscuits and cheese, six slabs of chocolate, a packet of salt, fifty cigarettes, tablet of soap, ½lb of sweets, a tin of dried milk, a packet of rolled oats, two 1lb tins of bacon and a tin of potatoes.

    Most Bevin Boys paid over their 22s 6d.

6. A left-hand bat and right-arm medium pacer, Gerald Smithson made his debut for Yorkshire in 1946 while still a Bevin Boy, when Yorkshire won the first post-war championship. He only made the West Indies tour because so many leading players, notably Hutton, Bedser, Compton and Edrich, chose to miss it (although Hutton went out later). Smithson played in the first two Tests, the second of which brought the 'Three Ws' – Walcott, Worrell and Weekes – together for the first time; he batted three times, scored seventy runs (highest thirty-five), averaging twenty-three. He didn't bowl.

    Smithson's career ended in 1956 (with Leicestershire from 1951). He died in 1970, aged forty-three.

7. Men were still being called up to do national service under wartime legislation, and until a new Act came into force on 1 January 1949, introducing peacetime conscription, which was to run until 1960, the government still kept coalmining open as an option. Only about 1,000 men a year took it up. The new Act exempted underground miners from call-up.

8. Dedicated to peace, friendship and global solidarity, the World Youth Festival is staged every two to four years. The first outside Europe was in Cuba in 1978. The sixteenth, in August 2005, in Venezuela, was attended by 20,000 from 170 nations.

9. *Muck, Silk and Socialism* (Paper Publishing, 2002).

Index